Table of Contents

BOOK 1 - SURVIVAL BASICS: THE ESSENTIALS YOU NEED TO KNOW 6

BOOK 2: WATER - THE ESSENTIAL ELEMENT OF SURVIVAL 20

BOOK 3: FIRST AID - MORE THAN A KIT, A LIFELINE 31

BOOK 4: YOUR CASTLE: DEFENSE AND SHELTER 43

BOOK 5: PROVISIONS FOR THE LONG HAUL - MASTERING FOOD SECURITY 56

BOOK 6: COMMUNICATING WHEN EVERYTHING ELSE FAILS 67

BOOK 7: HAM RADIO - THE VOICE OF HOPE 78

BOOK 8: SATELLITES AND DIGITAL TECH: THE FUTURE OF SURVIVAL 91

BOOK 9: BEYOND WORDS: NON-VERBAL COMMUNICATION AND COMMUNITY NETWORK. 103

BOOK 10: VHF AND UHF: BRIDGING DISTANCES WITH TECHNOLOGY 115

Dear Reader,

Welcome, and thank you profoundly for placing your trust in this book. You're embarking on a journey that transcends ordinary survival wisdom, diving deep into the knowledge and skills that could one day save your life and the lives of those you love. "Emergency Preparedness and Off-Grid Communication Bible for Preppers" is not just a book; it's a comprehensive guide designed to prepare you for the unpredictable, arming you with the essentials you need to thrive in any scenario.

This book is your ultimate guide in the world of prepping and survival, designed to be the equivalent of ten books condensed into one powerhouse resource. By "condensed," we mean that we've distilled the most essential and practical information from each critical discipline covered in this guide. We've stripped away the fluff, focusing solely on actionable knowledge because, in times of crisis, you need a manual you can turn to quickly — not a tome filled with unnecessary theory that won't help in real-life situations.

Each chapter in this book is crafted to provide you with concise yet comprehensive content that prepares you to handle a variety of challenges with confidence. Whether it's setting up off-grid communications, executing efficient survival gardening, or mastering the art of seed saving, you'll find that this guide cuts to the chase, delivering exactly what you need to know. This approach ensures that you can quickly assimilate and apply this knowledge when it matters most, making this not just a book, but a critical survival tool in your arsenal. This is why we market it as "10 books in 1" — because it covers the breadth and depth of ten separate guides, all in one streamlined, easy-to-navigate manual.

Within these pages, you'll uncover the fundamentals of survival, from securing an endless food supply and developing mental fortitude to mastering water procurement and purification. Each chapter is meticulously crafted to support you as you navigate through various survival challenges, whether it's creating a survival network, ensuring your family's safety, or unlocking the secrets to long-term food security.

Book Highlights:

- Essential Survival Skills: Learn innovative strategies for food, water, and shelter that go beyond the basics.
- First Aid Mastery: Delve into life-saving first aid techniques that do more than just scrape the surface.
- Defensive Measures: Discover how to fortify your home and protect your loved ones from potential threats.
- Communication Breakthroughs: Explore off-grid communication methods and the essential role of technology in survival.
- Community Building: Understand the importance of building a resilient community network for survival.

Recognizing the complex nature of survival in various environments and situations, this guide encompasses a broad spectrum of knowledge, from the essentials of first aid to advanced communication technologies and community networking. We use industry-standard terminology and practical advice to prepare you for any crisis you might face.

In the world of prepping, the notion that one can be a jack-of-all-trades is not just unrealistic; it's potentially hazardous. True preparedness isn't just about individual skills in food storage, medical aid, or self-defense; it's about weaving these skills into the fabric of a community that can collectively withstand the unexpected. This book, while covering the broad spectrum of essential prepping skills, focuses on a unique mechanism that could very well be the linchpin of survival: off-grid communication.

Let's face it, no single prepper can master every survival skill perfectly—some might be wizards with generators and solar panels, while others might have the greenest of thumbs for gardening or a sharp eye for security. However, the glue that holds these diverse skills together in a crisis is effective communication. When traditional systems fail, and the grid shuts down, it's the ability to communicate off-grid that can turn a group of isolated individuals into a resilient community.

This guide emphasizes off-grid communications as the critical skill set that maintains the cohesion of a prepper community. Why? Because in a crisis, the ability to swiftly share information and resources, coordinate responses, and reach out for help becomes the cornerstone of survival. It's about ensuring that when you're in a pinch, the call for help doesn't fall into silence but reaches those who can respond.

Imagine this scenario: a sudden natural disaster cuts off your small town from the surrounding areas. Power lines are down, cell towers are out, and the only way to coordinate community efforts efficiently is through the very skills this book prioritizes. Off-grid comms isn't just about keeping in touch; it's about maintaining operational unity when the rest of the world seems to crumble. It's the thread that stitches together individual capabilities into a quilt of communal resilience.

This book serves as your guide not just to surviving but to thriving through mutual support and shared strengths. Its unique focus on communication underscores a crucial truth in the prepping world: while alone you might move fast, together you'll go far. In survival, no prepper is an island; isolated, you risk disaster, but connected, you create a network of mutual survival and support. This isn't just a manual; it's a manifesto for building a community that stands strong against all odds.

To further enhance your readiness, we've included **three exclusive bonuses**:

#1: The Communication Signal Cheat Sheet - Master essential off-grid comms with top-secret tactics for Morse code, hand signals, radio codes, and whistle signals. Stay connected, command operations, and ensure safety without the grid. Essential knowledge for every prepper!

#2: Survival Checklist and Gear Guide - Equip yourself fully with our comprehensive gear guide. From critical survival tools to everyday necessities, ensure you're always ready. Dive deep into our detailed checklist to streamline your preparation for any scenario. Essential for every survivalist's arsenal!

#3: Survival Gardening and Seed Saving Guide - Delve into sustainable living with our detailed guide on survival gardening and seed saving. Learn to cultivate your food independence through practical advice on creating robust garden ecosystems and preserving seeds. Essential for any resilient prepper!

At the end of this book, you will find a QR code. Scanning this code grants you access to this specially curated set of resources designed to complement the information provided here. These additional materials are invaluable tools in your survival preparation arsenal, offering practical applications of the concepts discussed and furthering your ability to navigate emergencies effectively.

Your journey through these pages is very important to us. We've poured countless hours into ensuring that this book serves as your ultimate guide to survival. After diving into these secrets, should you find this guide instrumental in your preparation, I would be grateful if you could share your experience through a review. Your feedback not only supports our mission but also aids others in discovering the life-saving knowledge contained within.

Wishing you strength, wisdom, and preparedness on your path to mastering survival.

With warm regards,

Justn P. Brenner

Book 1 - Survival Basics: The Essentials You Need to Know

In a world where distressing headlines warn of foreign cyberattacks, potential grid failures, and looming threats to essential services, being prepared is no longer an option—it's a necessity. Whether you're a seasoned prepper or a curious survival enthusiast, this book is your roadmap to thriving in any scenario. Let's equip you with practical knowledge to face emergencies with confidence.

The first step in surviving any situation is to understand the environment you're in. Whether you're lost in the wilderness, navigating an urban disaster, or stranded in an unfamiliar area, knowing your surroundings is crucial. For instance, in the wilderness, recognizing which plants are edible or poisonous can make a significant difference. Similarly, in an urban disaster, knowing which buildings are safe to take shelter in can save your life.

One cannot overstate the importance of water in a survival situation. Our body can survive for several weeks without food but only a few days without drinking water. Locating a water source is your top priority. In the wilderness, look for streams or collect rainwater. In urban areas, water heaters or toilet tanks can be unexpected sources. Remember, purification is key; boiling water or using purification tablets can make questionable water safe to drink.

Prioritization saves lives. Remember the Rule of Three:

- 3 minutes without air
- 3 hours without shelter (in extreme conditions)
- 3 days without water
- 3 weeks without food

Focus on what's immediately important. For example, in harsh environments, shelter becomes your top priority after ensuring you can breathe.

Remember, Survival Basics isn't just theory—it's actionable steps for your safety.

In survival scenarios, keeping your cool is crucial. Panic can cloud your judgment and lead to very poor decisions. Instead, breathe deeply and evaluate your surroundings with a clear head. Building mental strength is key; practice by putting yourself in challenging but controlled environments, like camping solo. This prepares you to react calmly and wisely when under stress.

Being noticed is critical in rescue situations. Remember, three signals (like blasts of a whistle, fires, or flashes of light) universally mean you need help. Use anything at hand to make yourself more visible – mirrors, brightly colored clothing, or large symbols in an open area. Visibility could save your life.

Knowing first aid is more than just a survival skill—it's crucial for everyday situations too. Learn to treat cuts and scrapes, respond to hypothermia, and perform CPR. These abilities can save lives. Stay sharp by regularly going over your first aid skills and practicing them.

Flexibility is key when facing emergencies. The situation can change in an instant, and your success often depends on how quickly you can adapt. If you hit a roadblock, avoid getting stuck in annoyance. Look for alternative approaches instead. Quick thinking and the ability to bounce back are at the heart of many survival success stories.

The cornerstone of effective preparedness is regular practice. Engage in survival training, rehearse building shelters, practice water purification, and become intimately acquainted with your survival equipment. Frequent practice ingrains these skills into your muscle memory, ensuring you're well-prepared when the time comes.

Survival transcends mere physical readiness or the possession of the latest gadgets. It's grounded in deep knowledge, thorough preparation, and a resilient mindset. By mastering your environment, prioritizing tasks judiciously, maintaining your composure in stressful situations, and being ready to adapt, you arm yourself with the essential tools for any crisis. This guide is designed not just to educate but to empower you, giving you the confidence and skills needed to face unexpected challenges head-on.

Let's get into how to master your food resources. Ensuring a continuous food supply in survival situations might appear challenging, yet it's achievable with methodical strategies. This section is dedicated to guiding you through the process of establishing a consistent food source, no matter what your environment may present.

Your journey to an unending food source begins with understanding your environment. The natural world offers a rich array of edible plants, nuts, berries, and mushrooms. The key is identifying which ones are safe for consumption. A common example is the dandelion, which is not only widespread but also completely edible and packed with nutrients. By learning to identify and responsibly gather these natural foods, you essentially turn any outdoor setting into a resource-rich food market.

Protein is an essential part of your diet, especially in survival conditions. You can obtain this through effective fishing and trapping:

- Utilize basic but essential tools such as fishing lines, hooks, and traps.
- Acquire the skill of constructing and positioning traps, paying close attention to the behavior and trails of local wildlife.
- Place traps in high-traffic animal areas, particularly along paths or near bodies of water, to increase your chances of success.

A crucial aspect of fishing and trapping is the commitment to learn, apply, and adapt. Success comes from persistence and the willingness to modify techniques as you learn what works best in your specific surroundings.

Permaculture transforms survival from mere endurance to thriving sustainably, even in the harshest conditions. It's about forging a self-sustaining agricultural ecosystem that not only feeds you but also enhances your environment. Begin with manageable projects like nurturing sprouts or establishing a modest herb garden in your temporary shelter. These initial steps, though small, pave the way to a reliable and ongoing source of food, dramatically elevating your survival setup.

Permaculture goes beyond simple gardening; it is a philosophy that encourages a harmonious partnership with nature. By observing and replicating the natural world's patterns and relationships, you can cultivate a thriving ecosystem right where you are. Here's how:

Mini-Ecosystems: Even in constrained spaces, you can create a balanced micro-environment. Use the principle of companion planting—positioning complementary plants close together—to deter pests and boost soil health naturally, sidestepping the need for harmful chemicals.

In survival scenarios, efficient water use becomes critical. Adopting permaculture principles means designing your growing space to maximize water efficiency:

Implement strategies like mulching, constructing swales, and harvesting rainwater to ensure your plants remain well-watered with minimal resources.

A Basic Rainwater Capture System.

1. Collecting rainwater can be a simple yet effective way to conserve water, particularly in survival situations. Here's how to set up a basic rainwater capture system:
2. Choose Your Containers: Start with clean, sturdy containers like barrels, buckets, or large jugs. Ensure they're made from food-grade plastic or coated metal to avoid contaminating the water.
3. Find the Right Location: Place your containers under downspouts or gutters where rainwater runs off your roof. If you don't have gutters, position them in open areas where they can catch direct rainfall.
4. Prepare the Catchment Area: If using a roof, ensure it's free from heavy debris, animal droppings, or chemical residues. Clean your gutters and downspouts to prevent blockages and ensure clean water flow.
5. Set Up a Filtration System: Cover the opening of your containers with a fine mesh or screen to filter out leaves, insects, and other debris. This step is crucial for maintaining water quality.
6. Secure Your Containers: Make sure your containers are stable and won't tip over in strong winds. Secure them to a solid structure or weigh them down.
7. Connect Multiple Containers: If you have several containers, connect them with pipes or hoses so when one fills up, the excess water flows to the next container. This increases your

collection capacity.

8. Maintain Water Quality: Add a few drops of non-scented bleach to each gallon of collected water to prevent algae and bacteria growth. Always filter or boil the water before using it for drinking or cooking.

Regular Maintenance: Check and clean your containers, filters, and catchment area regularly to ensure the system remains effective and the water stays clean.

By implementing this basic rainwater capture system, you can significantly increase your water reserves, essential for your garden's survival during dry spells and crucial for your own needs in emergencies.

Remember, the essence of permaculture in survival is not just about sustaining life but improving your living conditions by working in tandem with the environment. By adopting these practices, you not only secure a consistent food supply but also contribute to a healthier, more sustainable way of living, regardless of the circumstances.

Getting your survival garden to a state of flourishing begins right from the ground up – the soil is your foundation. In the realm of permaculture, soil health is king. That means getting down and dirty with composting, transforming your everyday kitchen leftovers and garden refuse into gold for your plants. Set up a basic compost pile or bin, and watch as what was once waste turns into a treasure trove of nutrients, giving your garden's soil the boost it needs to support strong, resilient plant growth, even when times get tough.

Now, let's talk about transforming your plot into something out of a survivalist's handbook – an edible forest garden. Think of it as an ecosystem on your doorstep, integrating trees, bushes, perennials, and annuals into a single, self-feeding food production zone. It's about creating a mini-nature reserve that feeds you. Yes, kicking off this venture takes effort, but envision the end game: a vibrant, self-sustaining garden that's less about maintenance and more about harvest.

But here's the kicker: survival gardening is all about staying fluid. You've got to roll with the punches – changing weather, shifting soil conditions, uninvited pests – and adapt. Mix things up with crop rotation, bring in new plant soldiers to combat old challenges, or rejig your garden's game plan to meet current needs. It's this kind of flexibility that turns a static garden into a living, breathing survival machine. By staying adaptable, you ensure that come rain or drought, your garden continues to be a beacon of food security, proving time and again that the ability to adapt is not merely about surviving; it's about thriving against all odds. When it comes to foraging, the name of the game is safety. Identifying the right plants is critical, as some can nourish you while others can harm you. Always double-check the plants you plan to eat; never go by just looks alone. Remember, plants with milky sap or groups of three leaves can be bad news, signaling danger. And a bitter taste? That's nature's way of saying, "Don't eat this." Stick to these guidelines to avoid turning a meal into a mishap.

Educate Yourself: Before you start, learn about the local flora. Use reliable guides or apps, attend workshops, and consider consulting with experienced foragers.

- Safe and Responsible Foraging: Guidelines for Gathering Wild Edibles
- Avoid Harmful Plants: Stay away from plants with milky sap, plants with groups of three leaves, or those with an almond scent in the wood or leaves, as these features often indicate toxicity.
- Forage in Safe Areas: Avoid foraging near busy roads, industrial areas, or places that might be contaminated with pesticides or herbicides.
- Harvest Sustainably: Only take what you need and leave enough behind for the plant to continue to grow. Never uproot or destroy entire plants unless you're harvesting tubers.
- Use All Your Senses: Observe the plant's color, texture, and smell. However, even if a plant looks and smells pleasant, it doesn't guarantee it's safe to eat.
- Be Cautious with Mushrooms: Mushrooms can be particularly dangerous due to their potent toxins. If you're not 100% sure of a mushroom's identity, don't eat it.
- Respect Wildlife and the Environment: Remember, you're sharing the space with wildlife who rely on these plants for their survival. Always forage responsibly.

By following these basic tips, you can enjoy the bounty of nature while minimizing risks, turning nature's offerings into a safe, sustainable food source during emergencies.

Preserving Food.

Learn some essential food preservation techniques. Long-lasting food storage is crucial, especially when you're counting on your supplies to get you through tough times. But it's not just about keeping food from going bad; it's about maintaining its goodness. Here's a technique that's stood the test of time:

Smoking: This isn't just for flavor. Smoking food is a tried-and-true method to extend shelf life. By exposing food to smoke from burning wood, you can preserve meats and fish for months. But there's a knack to it: you need to control the smoke and heat just right. The aim? Dry out the food without cooking it completely. Get this balance right, and you'll have tasty, preserved food ready when you need it most. This method not only adds a delicious smoky flavor but also keeps your food safe for longer periods, ensuring you have a reliable food source whatever the situation.

Follow these steps to smoke your food successfully:

1. Preparation: Start by preparing your meat or fish. Clean it thoroughly, then salt it or use a brine solution to help in the curing process. This step is crucial for flavor and preservation.
2. Choosing Wood: Different woods impart different flavors. Hickory and mesquite offer strong tastes, while apple and cherry woods provide a milder, sweeter flavor. Avoid resinous woods like pine, as they can spoil the food.

3. Setting Up Your Smoker: You can use a commercial smoker or make a DIY version using a barrel or an old fridge. Ensure there's a space for the fire, a rack to place the food, and good ventilation for smoke circulation.
4. Controlling Temperature: Ideal smoking temperatures are between 200-250°F. Use a thermometer to monitor. Too hot and you'll cook the food instead of smoking it.
5. Applying Smoke: Once the smoker is at the right temperature, add your wood. Then, place the food on the racks and close the smoker. The key here is low and slow; rushing can lead to unevenly smoked or overly dry food.
6. Timing: Smoking time varies based on the type of food and its size. Fish might only need a few hours, while thick cuts of meat could take all day. Check periodically, but remember every time you open the smoker, you lose heat and smoke.
7. Storage: Once smoked, let the food cool. Then, wrap it in butcher paper or vacuum seal it. Store in a cool, dry place. Properly smoked food can last for weeks, extending your food resources during emergencies.

By mastering smoking, you not only enhance the flavor of your food but also increase its shelf life, a vital skill for any survival situation.

Drying, or dehydrating, stands out as the most straightforward, low-energy food preservation method. Removing moisture stops spoilage-inducing microbes in their tracks. This method works well for a variety of foods including fruits, vegetables, meats, and herbs, extending their usability for months, sometimes years. In sunny, arid conditions, sun drying is effective; however, in damp areas, using an oven or dehydrator is recommended. Foods preserved this way maintain most of their nutritional value and can be easily rehydrated, making them perfect for emergency storage.

Salting isn't just for enhancing flavor—it's a powerful preservation technique for meats and fish. It works by drawing out moisture and creating a bacteria-unfriendly environment. The two primary approaches are dry salting, where food is completely covered in salt, and brining, involving submersion in a saltwater solution. Both methods take time and must be done under suitable conditions. When executed well, salting can keep food safe to eat for extended periods. For even longer storage, consider combining salting with drying or smoking.

Fermentation is more than preservation; it's a way to boost food's nutritional and flavor profiles. This method promotes beneficial bacteria and yeasts, turning basic ingredients into nutrient-rich foods like sauerkraut from cabbage or yogurt from milk. Properly fermented items can last for months and are packed with vitamins, probiotics, and enzymes. Starting with fermentation is easy and requires minimal equipment, offering a unique, nutritious addition to your survival food stockpile.

Just to give you an idea of how straightforward it is, here's the method for creating your own sauerkraut, a fermented food full of probiotics:

Sauerkraut Recipe

Ingredients:

- 1 medium cabbage (about 2 pounds)
- 1-3 tablespoons of non-iodized salt (sea salt is ideal)

Method:

1. Prepare the Cabbage: Remove the outer leaves of the cabbage and save a couple of the cleaner ones. Slice the rest of the cabbage thinly.
2. Salt and Squeeze: Place the shredded cabbage in a large bowl. Sprinkle with salt. Start massaging the salt into the cabbage. Continue squeezing and crushing it until there's enough liquid (brine) to cover the cabbage in the bowl.
3. Pack the Jar: Pack the cabbage tightly into a clean, airtight jar (a mason jar works well). Press down firmly to ensure the cabbage is submerged in its own brine. Leave some space at the top to allow for expansion.
4. Seal and Store: Place one of the reserved whole leaves on top to keep the shredded pieces submerged. Close the jar tightly. Store it at room temperature, away from direct sunlight, in a place where you can check it regularly.
5. Burp the Jar: For the first few days, open the jar once a day to release gases produced during fermentation. This is known as "burping" the jar.
6. Taste and Wait: Start tasting the sauerkraut after a few days. Once it reaches a flavor and acidity level that's to your liking (typically between 5 to 14 days), transfer it to the refrigerator to slow fermentation.

Remember, the key to successful fermentation is keeping the cabbage submerged under the brine. This method can be applied to various vegetables, so feel free to experiment once you're comfortable with the basics. Enjoy your homemade sauerkraut as a side dish, in sandwiches, or as a flavorful addition to salads!

Canning is your go-to method for long-term food storage. It involves sealing foods in jars and heating them to a temperature that eliminates harmful bacteria and enzymes. You can can almost anything – fruits, veggies, meats, and even full meals like soups.

Yes, canning needs a bit of upfront investment in gear and learning. But get it right, and you've got food that stays safe, tasty, and nutritious for years.

Incorporate canning and other preservation methods into your survival plan to secure a constant food supply. This isn't just about staying fed; it's about keeping meals interesting and nutritionally balanced, which is crucial when times get tough.

Start practicing these techniques now, not when crisis hits. This way, you can iron out the kinks and figure out what works best for you. Plus, spreading this knowledge in your community builds a stronger, more resilient group, ready to face whatever comes.

Mastering the art and science of food preservation links you to centuries of human ingenuity and resilience. It's more than survival; it's about maintaining a standard of living and health, ensuring that, come what may, you and yours can keep thriving.

Gardening Techniques.

Why don't you get stuck into learning some revolutionary gardening techniques. In the pursuit of self-reliance, particularly in off-grid living, innovative gardening methods such as hydroponics and aquaponics stand out as game-changers. These systems transform traditional food cultivation practices, enabling you to grow vegetables and rear fish simultaneously in a self-sustaining loop. This dual approach not only secures a continuous food supply but also introduces a novel concept in self-sufficient living.

These gardening methods are a beacon of hope, especially in environments with limited space or degraded soil. By adopting these techniques, you can overcome conventional farming challenges and pave the way for a more resilient and efficient food production system.

Hydroponics is a specific technique of cultivating plants without soil. It relies on a water-based solution enriched with essential minerals. This method offers you precise control over your plants' nutrients, pH levels, and moisture, leading to quicker growth and larger yields than typical soil-based agriculture.

Here's how to get started:

Construct a basic hydroponic setup using readily available materials such as PVC pipes, a water pump, and planting containers.

Opt for fast-growing, leafy vegetables like lettuce, spinach, and various herbs, which are particularly well-suited for hydroponic systems.

Expect to harvest your crops in just a few weeks after planting, thanks to the efficient nutrient delivery system.

One significant benefit of hydroponics is its water-saving prowess, using up to 90% less water than traditional methods. This feature makes it invaluable in dry regions or when water conservation is paramount. Moreover, the direct delivery of nutrients to the plants eliminates the need for chemical fertilizers, promoting a purer, more sustainable form of farming.

Aquaponics elevates the hydroponics concept by integrating fish into the cycle. In this ecosystem, fish waste serves as a natural nutrient source for plants, while the plants clean and purify the water, which is then recirculated to the fish tanks.

Setting up an aquaponics system is a bit more complex than hydroponics but offers greater rewards:

- Begin with setting up tanks for your fish and grow beds for your plants.
- Choose resilient fish species like tilapia or catfish for your system, though decorative fish like goldfish are also suitable.
- Enjoy the dual benefits of harvesting both fresh produce and protein, streamlining your food production process.

Aquaponics systems can be installed indoors or outdoors, offering versatility based on your spatial and climatic limitations. By integrating these innovative gardening methods into your off-grid living strategy, you're not just growing food; you're cultivating resilience, efficiency, and a deeper connection with your environment. Embrace these techniques to revolutionize your approach to food self-sufficiency.

In the concrete jungle of urban settings or in confined living conditions, harnessing the power of hydroponics and aquaponics becomes a transformative strategy. Through vertical gardening, you can elevate your farming vertically, allowing layers of productivity to thrive within a compact space. This inventive approach enables you to transform even the tiniest balcony, garage, or indoor room into a verdant and flourishing garden.

By leveraging vertical space, these systems defy traditional gardening limitations, offering a scalable solution that fits your available area. In embracing these innovative techniques, you contribute to a sustainable future, ensuring you and your community can withstand any challenge with resilience and a proactive attitude.

Making use of waste.

The mantra "Waste not, want not" is your guiding principle in survival scenarios. Here's how to apply this approach effectively:

Animal by-products: Transform bones into nourishing broth, use hides for crafting shelter or clothing, and consider other parts for various needs.

Plant scraps: Don't discard plant remnants; instead, turn them into compost to enrich your garden soil, closing the loop in your food production cycle.

This comprehensive approach to resource usage not only minimizes waste but also amplifies the sustainability of your survival practices, ensuring every available resource is utilized to its fullest potential.

The strength of community becomes your backbone in times of need. Here's how to cultivate and leverage communal bonds:

Share resources, knowledge, and efforts with neighbors and fellow community members to enhance collective food security.

Initiate or participate in a barter system, exchanging goods and services, thereby reducing the dependence on traditional currency systems.

Fostering a collaborative environment not only diversifies and stabilizes the food supply but also builds enduring relationships and a support network crucial for survival and prosperity.

Surviving and thriving, particularly in adverse conditions, demand innovation, wisdom, and the willingness to adapt. By harnessing the environment and community around you, and by employing sustainable practices in fishing, trapping, and gardening, you lay the foundation for a secure and sustainable food source. This strategic approach not only ensures your physical well-being but also fosters a sense of stability and continuity during tumultuous times, empowering you to face challenges with confidence and peace of mind.

Developing Mental Fortitude.

Mental toughness isn't just a bonus; it's as vital as the knife in your pocket or the fire-starter in your pack. It's the backbone of your survival spirit, turning potential breakdowns into breakthroughs. Grasping the power of mental fortitude transforms the way we tackle survival from the ground up. Think of your mental space as your ultimate survival kit, stocked not with gear, but with skills to beat stress, embrace change, and maintain an unwavering positive attitude. These aren't just ideas; they're the mental equivalents of a well-honed blade or a trusty compass, ready to navigate you through the harshest conditions.

Unchecked stress is like fog on a battlefield; it muddies your thoughts and misguides your steps. Learning to manage it isn't a luxury—it's as necessary as finding clean water in the desert. Tools like deep breathing, meditation, or even structured problem-solving can clear your mind, acting as your internal compass recalibrating. These aren't fluffy concepts but proven strategies that lower your body's stress responses, making you more alert, more decisive, and more capable when every decision counts. Make these techniques part of your everyday life, and watch as they turn you into a calmer, more resilient survivor, someone ready to face whatever the wild throws at you with a clear head and a steady hand.

By embedding these mental practices into your daily routine, you're not just surviving; you're constructing a fortress of calm in your mind, ready to withstand any storm. This mental preparedness isn't just about getting through the night; it's about emerging from the other side stronger, more capable, and with a clear vision for the path ahead.Embodying Bruce Lee's philosophy to "Be water, my friend," highlights the importance of adaptability – a quality as vital as a reliable shelter during a storm. Introducing small changes into your daily life, such as varying your routine or embracing new experiences, builds your capacity for significant life alterations. This mental agility ensures you remain unyielded, flowing seamlessly through the adversities life presents.

Maintaining a positive outlook illuminates your journey, guiding you through the storm with the radiance of hope and perseverance. Positivity, underscored by research linking optimism to improved health, acts as your psychological bulwark, nurturing your spirit through dark times. The act of recording daily gratitudes, such as the serenity of nature or the comfort of a safe haven, shifts your perspective from the burdens you carry to the blessings you harbor, reinforcing your mental and emotional fortitude for the trials ahead. This relentless spirit, cultivated through gratitude and optimism, transforms the wilderness of despair into a landscape of hope.

Strategic thinking is your mental compass in uncharted territories. It's not merely a response mechanism; it's your ability to sculpt the future from the fabric of the present. Begin with clear, tangible objectives. In the wild, this could translate to mapping out your trail, allocating your resources wisely, or establishing daily goals. Segmenting daunting tasks into smaller, achievable milestones fosters a sense of mastery and direction. Example: If stranded in a remote location, identify high ground for signaling, estimate your food supplies for weekly consumption, and plan for shelter improvements daily.

Practicing mental drills is like running a survival scenario in your head. Picture different tough spots you might land in and work out how you'd wiggle out of them. This method isn't just for athletes aiming for gold; it's for you to mentally leap over survival challenges. This kind of prep work cuts down on fear and boosts your confidence, giving you a mental playbook when things get real.

Keeping your emotions in check is your lifeline in the chaos of survival. The wild can whip up a storm of feelings like fear and frustration. Recognizing these feelings and understanding they're just passing clouds can help you stay clear-headed. Use the "STOP" method: Stop, Take a breath, Observe your thoughts, and Proceed with a plan. This helps you stay level-headed and make smart choices.

Building a tough-as-nails mindset is like armoring up your inner self. It's about getting better at dealing with hard times, adapting to changes, and not giving up when the going gets tough. Pushing your limits, picking up new skills, and stepping out of your comfort zone all help build this mental toughness and confidence.

Your mind needs to be as ready as your survival kit. By getting a grip on stress, being flexible, staying positive, planning smart, and staying calm, you're all set for whatever comes your way. Survival is more than just making it through; it's about coming out on the other side smarter, stronger, and ready for the next challenge.

Remember, there's real power in numbers when it comes to survival so build your survival network. The idea of the lone wolf making it on their own is more myth than reality. Building a strong network of reliable folks can seriously up your chances when things go south.

Think of your network like a strong spider's web, where each connection adds to its strength.

You're at the center, linked up with family, friends, neighbors, and even folks you haven't met yet who are also geared up to handle tough times.

Why a network matters:

1. Pooling resources means everyone has more to work with.
2. Sharing skills means everyone has more know-how.
3. Having each other's backs means everyone feels stronger and less stressed.

Start with the people closest to you or your family and then branch out. Look for folks who bring something different to the table - someone who's a whiz with first aid, another who knows all about local plants, or someone else who can fix just about anything.

Think about setting up a local skill-swap meetup - you could all come out knowing a lot more about staying safe and looking out for each other.

Trust is the glue that holds your network together. Be the person others can count on, and don't be shy about getting involved in community stuff like clean-up days or prep workshops. It's a great way to spot other solid folks who take survival seriously.

Keeping the lines of communication clear is non-negotiable. Sure, our phones and computers are great day-to-day, but what if they go down? Have backup plans like two-way radios or other signal methods, and make sure everyone knows how to use them.

1. Always keep your gear in working order.
2. Get comfortable with old-school ways like radios.
3. Set up regular times to touch base if digital fails.
4. Make sure everyone knows basic signals or signs.
5. Keep an up-to-date list of contacts and essential info where you can all find it.

By teaching each other, making clear plans on how to help out in a pinch, and keeping everyone looped in, your group won't just be ready to face tough times - you'll be setting the standard for what a tight-knit, ready-for-anything community looks like. In a world full of unknowns, this network is your tribe, ready to face the wilderness of the modern world together.

Teaching kids about survival.

Teaching kids how to survive isn't just about the cool stuff like making fires or finding your way in the wild without a compass. It's also about teaching them to be tough, smart, and to look out for others. Start with the basics: show them how to find and purify water, why staying hydrated is key, how to whip up a decent meal with whatever's around, how to build shelters that'll keep them safe in different settings, and what to do if they ever get separated from their group.

Make survival skills a part of their everyday life. Turn family camping trips into fun, hands-on lessons where they get to practice making a fire, setting up camp, and finding their way with a map and compass. Get them involved in the garden to teach them where food comes from and why it's important to take care of the earth. Use everyday moments, like a change in the weather or a scraped knee, to teach them about nature and first aid.

Push them to think for themselves and come up with solutions when they hit a snag. Could be figuring out different ways to use a stick or coming up with a plan when their fort collapses. It's about making them feel like they can handle anything that comes their way and boosting their confidence along the ride.

Don't forget to teach them about taking care of our planet. Explain how everything in nature is connected, why we need to protect it, and how they can do their part. Take them on hikes, join in on local clean-up days — anything that helps them feel connected to the great outdoors.

Case Study: Take the story of a kid who used what they learned to get through a surprise storm on a hiking trip. They found shelter, stayed hydrated, and signaled for help just like they'd been taught. It wasn't just a story of survival, but a real-life lesson on why knowing this stuff matters. It shows that these aren't just skills for "one day" — they're for real life, right now, and they can make a huge difference.

By integrating these lessons and principles, you're not just preparing them for potential survival scenarios; you're equipping them with skills and mindsets that will serve them in all areas of life. The goal is to foster a generation that's not only capable of surviving but thriving, no matter the circumstances they face.

Teaching kids about survival isn't just about showing them how to light a fire or find their way without a GPS. It's about teaching them to work well with others, to think on their feet, and to care about their community and the environment.

Start with the basics of teamwork by getting them involved in team sports or group projects. Show them how everyone has something unique to offer and how working together can make things easier and more fun.

Bring history into it by telling them about explorers who survived tough situations because they worked as a team. These stories aren't just cool tales from the past; they show how important it is to stick together and use everyone's skills.

Encourage their curiosity. If they ask about the stars, a weird bug, or how people in different parts of the world live, dive into those questions with them. It's a chance to learn together and show them that understanding the world is a big part of being prepared.

Lastly, focus on building their confidence. Celebrate the small stuff they do well, and give them chances to learn from mistakes. Confidence comes from knowing you've got skills and can handle whatever comes your way.

Remember, it's not just about surviving; it's about teaching them to live responsibly and care for others. They'll carry these lessons with them forever, making them ready for anything and good people to boot.

BOOK 2: Water - The Essential Element of Survival

Finding a reliable source of water is your number one priority in any survival situation. Your body depends on water for survival far more urgently than it does for food, as it's vital for temperature regulation, nutrient transportation, and toxin elimination.

In different environments, your approach to finding water will vary. While rivers, lakes, and streams are the most obvious places to start, don't overlook other sources like rainwater, dew, or plant moisture. Knowing where to find water and how to make it safe for drinking is critical.

In the wilderness, look for signs of water by observing the vegetation and wildlife. Greener, lush areas usually signal nearby water. Follow animal tracks as they can lead to natural water sources. Also, insects like mosquitoes gather near water, and birds typically head to and from water sources at dawn and dusk. Always remember, moving downhill is a good strategy in hilly or mountainous areas as water naturally flows downwards.

Imagine you're stranded in a mixed environment with forests, hills, and a few clearings. The first thing you'd do, after ensuring your immediate safety, is to scout for water. Start by scanning the area for lush greenery—a hint of moisture in the environment. Remember, your observations now can mean the difference between resilience and desperation.

As you walk, notice the direction birds are flying, particularly in the early morning or late evening. They're heading to their water source for the day. Tracking these natural guides can lead directly to your next drink. But don't rush. Approach the water source cautiously, observing if there are any animal tracks leading to or from the site. These tracks can guide you to water but also warn of potential predators or contaminated areas.

Now, suppose you encounter a rocky terrain that's directing you downhill. Your instincts and knowledge tell you water follows gravity, leading to higher chances of finding a stream or pond at lower elevations. As you move, the sound of flowing water begins to reach your ears—a sign that your strategy is paying off.

Upon finding a water source, your next task is assessing its safety and beginning the purification process. But let's remember, this water, while a sight for sore eyes, isn't ready to drink yet. Your next steps involve filtering and boiling, ensuring the elimination of pathogens.

Rainwater.

Rainwater can be a lifesaver, especially when other sources are not available. You can collect rain efficiently using items like tarps, ponchos, or large leaves to funnel it into your containers. In drier climates, even small amounts of rain should be captured to maximize your water reserves.

Let's expand on rainwater collection. Imagine clouds are gathering, and you've found a relatively open space. You set up a tarp with its corners tied to trees and the center slightly weighed down. As rain falls, water collects at the center and funnels into your storage container—a clean, improvised canteen or even a boiled-out plastic bottle. You've turned a simple piece of survival gear into an efficient water collection system.

But, rainfall is unpredictable. So, during dry spells, your attention turns to morning dew, nature's small but significant bounty. Using cloth or even your own clothes, you wipe the dew from leaves, grass, and any surfaces you find, squeezing the gathered moisture into your mouth or container. It's slow, meticulous work, but it's a silent testament to the resourcefulness demanded by survival.

Using plants for hydration.

It's a good idea to understand how to use tree sap. In forested areas, certain trees can provide life-sustaining liquids. Birch and maple trees, for example, contain sap that can be consumed for hydration. The early spring is the best time for sap flow. To tap into this resource, you need to make a small, diagonal incision in the tree's bark and insert a small tube or hollow stick to guide the sap out—always doing so responsibly to avoid harming the tree. Collect the sap in a container and boil it to reduce any pathogens. While less common than finding a stream or collecting rain, tapping trees for water is an age-old method that requires both respect for nature and survival knowledge.

In every scenario, being prepared and informed transforms challenging situations into opportunities for survival. By expanding your knowledge base and practicing these techniques, you ensure that when faced with the unexpected, you're not just reacting; you're responding with calculated, educated decisions that increase your chances of enduring survival situations. This proactive approach is what defines a truly prepared survivor.

Let's look at cacti as a water source. Cacti can store significant amounts of water in their flesh, making them potential lifesavers in desert environments. However, not all cacti are safe to use, and some can be downright harmful. The key to utilizing cacti safely involves:

Identification: Familiarize yourself with the types of cacti that are safe to consume. The Barrel cactus, for example, often contains drinkable water. Avoid the Saguaro and the Prickly Pear cactus, as these can contain compounds harmful to humans.

Harvesting: Approach with caution. Use gloves or a piece of cloth to protect your hands from the spines. A sharp knife or a cutting tool is essential for piercing the tough skin of the cactus. Cut a small section to begin with, as this minimizes damage to the plant and preserves it for future use.

Extraction: Once you have safely opened the cactus, you can either scoop out the flesh and press it to extract the water or, if the flesh is soft enough, suck directly on it. Be cautious, as the liquid may need to be strained through a cloth to remove any solid particles or harmful substances.

Filtering: While cactus water is often less contaminated than other natural sources, it's prudent to filter it through a clean cloth. This can remove particulate matter and some impurities, making it safer for consumption.

Not All Water is Equal: The water stored in cacti can vary in taste and quality. Some may find it slightly bitter, but in survival situations, the priority is hydration, not flavor.

Conservation: If you find a water-bearing cactus, use it sparingly. Preserve the plant for future use or for others who might find themselves in dire straits. Cutting off only what you need and covering the exposed flesh with mud or clay can help the cactus to heal and continue storing water.

Beyond cacti, many other desert plants can provide moisture. The Yucca plant, for instance, can hold water in its roots and trunk. Similarly, the Agave plant can offer moisture and sap that's rich in water. This knowledge, coupled with a well-rounded approach to sourcing and purifying water, ensures that you're ready to sustain yourself in a variety of environments, reinforcing the importance of water to survival and underscoring the need for preparedness and resilience in the face of adversity. However, be aware that some plants can be misleading and dangerous. Only use plants as a water source if you're certain they're safe.

Other plants, like birches, offer sap which can be a hydration source, and collecting dew from leaves or tree crevices can also provide water. Educate yourself on local plants and water collection techniques before you face a survival situation. Survival courses, botanical guides, and advice from experienced outdoorspeople are invaluable resources. In critical situations, knowing how to safely use the natural resources available can save your life.

Remember, while plants can provide emergency hydration, they're often a last resort. Always prioritize finding and purifying water from more conventional sources first. Techniques like using transpiration bags on leafy branches can supplement your water supply in difficult environments. This slow but steady method works well when other sources are scarce.

Understanding and practicing these methods will prepare you for maintaining hydration in any situation, reinforcing the essential role of water in survival.

Making Use of Dew.

Think about utilizing dew. When traditional water sources are scarce, every drop counts, especially in survival situations where conventional methods aren't an option. Let's dive into unconventional yet effective strategies like dew collection, a technique as old as nature itself, yet remarkably efficient for the savvy survivor.

The world is your water bottle when you know how to collect dew. Start your day before the sun does or as twilight fades, armed with absorbent materials — think clean towels, cotton strips, or even dense grass clumps. Purity is paramount here; any contamination defeats the purpose.

Secure these materials around your legs; makeshift gaiters soaked in morning dew can be your unexpected oasis. Trek through meadows or lawns where dew gathers like nature's bounty. It's a silent, serene collection method, placing you right at the heart of nature's lifeline.

Extracting this precious moisture requires patience. Once your materials are heavy with dew, wring them out over a vessel, every drop a triumph. This might test your patience, but in survival, patience is more than a virtue; it's your lifeline.

However, dew, like all water in the wild, hides unseen dangers. Boiling is your best friend here, annihilating pathogens and making your hard-won water safe. If fire is a luxury, then purification tablets or solar methods are your go-to. Remember, clarity doesn't equal purity.

Transforming Snow and Ice into Safe Water.

In the grip of winter, snow and ice are your dormant water reserves. But consuming them as is can chill your core, a dangerous trade-off. Melt them first, away from the deceptive warmth of your mouth, preserving your body heat.

When gathering snow or ice, always opt for the freshest supply you can find. Fresh snow is lighter and cleaner compared to compact ice or older snow layers that may have been exposed to contaminants over time. Use clean tools, such as a stainless-steel pot or a plastic bag, to collect your snow or ice to avoid any direct contact with your hands which may introduce bacteria.

Melting snow or ice can be done using a portable stove, a campfire, or even the body heat from your own body if other resources are scarce. However, direct consumption of cold snow can lead to hypothermia. If using a fire or stove, place the snow or ice in a pot and allow it to slowly turn into water. Don't fill the pot to the brim with snow; start with a small amount and add more as it melts to prevent scorching your container and to ensure even melting.

Melting snow and ice can be energy-intensive. To conserve fuel, mix snow with existing liquid water (if available) to speed up the melting process. Covering the pot with a lid retains heat and accelerates melting. Remember, one pot of snow can yield significantly less water than expected, so plan accordingly.

In cold environments, dehydration might not be as noticeable, yet it's equally dangerous. Monitor your hydration levels and consume melted snow regularly. Don't wait until you're thirsty to start melting snow; by then, you're already dehydrated.

Purification Tips:

- Boiling: After melting, bring the water to a rolling boil for at least one minute to kill any pathogens. At higher altitudes, extend this to three minutes due to the lower boiling point of water.
- Chemical Treatment: If boiling isn't an option, use chemical purification methods like io-

dine or chlorine tablets. Follow the instructions carefully, as cold water may require a longer treatment time.
- Filtering: Pre-filter melted snow through a clean cloth or coffee filter to remove any particulate matter. Post-treatment, use a portable water filter to remove any remaining microorganisms.
- Taste Adjustment: Melted snow can taste flat due to the lack of oxygen. Improve the taste by shaking it in a container or pouring it back and forth between two containers to aerate it.
- Avoid Polluted Areas: Steer clear of snow near roads, industrial areas, or yellowed snow which can contain harmful pollutants or animal waste. Always source snow from clean, untouched areas.
- Observation: Pay attention to the surrounding environment. Signs of animal activity, traces of vegetation, or human impact can indicate compromised snow.

By following these detailed steps, you transform frozen resources into life-sustaining water. This methodical approach not only maximizes your safety but also ensures that you remain hydrated and healthy in harsh winter conditions. Remember, in a survival situation, your ingenuity and knowledge are as crucial as the gear you carry.

Water Purification.

Beyond the basics lies the art of making water drinkable. Boiling is the gold standard, a rolling boil the barrier between you and illness. But what if fire eludes you? Chemical tablets, used with precision according to the instructions, can save the day. The sun, too, offers purification powers, turning a clear bottle into a UV sanctum over six painstaking hours.

Never underestimate the threats lurking in untreated water. Your diligence here is a barrier against the invisible enemies waiting to weaken you.

Conservation isn't just a principle; in survival scenarios, it's as critical as finding water itself. Avoid unnecessary exertion under the scorching sun, preserving sweat like the precious resource it is. Dress smart — lightweight, breathable fabrics can be the difference between dehydration and endurance.

Every action in a survival situation should be measured, every drop of water accounted for. Prioritize hydration above all else; cleanliness can wait. This isn't about comfort; it's about survival.

Adapting to Your Environment: Survival is not one-size-fits-all; it's about adapting to your surroundings. Whether it's collecting rainwater in a forest or scooping up snow in a blizzard, your environment dictates your strategy. Look around, think creatively, and use what nature provides.

Practice these techniques before you ever need them. Familiarity breeds confidence. Turn every camping trip into a rehearsal for survival, where every challenge is an opportunity to refine your skills.

Survival is personal, but it's also communal. Sharing your knowledge, from water collection to purification, strengthens not just your chances but those of your community. Organize workshops, lead by example, and build a network of prepared individuals.

Your experiences, shared freely, transform individual knowledge into collective power. Teach your family, engage your friends, and together, build a community where everyone is equipped, informed, and ready.

Understanding the crucial role of water purification is the first step. Untreated water from natural sources can carry harmful organisms and chemicals. Proper purification eliminates these risks, making water safe for consumption.

Boiling is the gold standard for water purification. Bring water to a rolling boil for at least one minute, longer at higher altitudes, to eradicate common pathogens. Though straightforward, boiling requires a heat source and patience for the water to cool afterward.

Using regular bleach can effectively purify water. Typically, add eight tiny drops per gallon, mix, and wait 30 minutes This method combats many pathogens, although it's not universally effective against all viruses and parasites.

Water filters range from simple straw versions to advanced pump models, trapping harmful microorganisms. The effectiveness of a filter is determined by its pore size, with smaller sizes capturing more contaminants. Regular maintenance is essential for optimal performance.

Portable UV purifiers offer a modern solution by neutralizing microbes through DNA disruption. Ideal for clear water, these devices need power to operate and perform best in non-turbid conditions to ensure all pathogens are exposed to UV light.

This method leverages sunlight to purify water. Fill clear plastic bottles and expose them to direct sunlight for six hours, or two days if cloudy. The combination of UV rays and heat effectively neutralizes harmful organisms, making it a cost-effective and simple method, though its success depends heavily on weather conditions.

Distillation is a thorough method for purifying water, involving the evaporation of water and subsequent condensation to separate it from contaminants. Here's a step-by-step guide to creating a simple, effective homemade distillation setup:

You'll need two containers (one larger than the other), a heat source, a concave lid or a large bowl, and a small weight. Choose materials that can withstand boiling temperatures without releasing harmful chemicals.

Fill the larger container with the water you need to purify. Place the smaller container inside—this is where the distilled water will collect. Ensure the smaller container is elevated or floated so it doesn't tip over.

Place the concave lid or bowl on top of the larger container. The concave shape ensures that condensed water drips into the smaller container. To improve efficiency, fill the lid or bowl with cold water or ice. This cold surface accelerates condensation.

Place the setup on your heat source and bring the water to a boil. When the water reaches its boiling point it will turn into steam and rise. When the steam hits the cold lid or bowl above, it will condense back into liquid form, free from the contaminants that remain in the larger container.

The condensed water dripping into the smaller container is distilled and safe to drink. Regularly replace the cold water on the lid or bowl to maintain the efficiency of condensation.

Continue boiling and condensing until you have enough distilled water for your needs. Remember, the process does remove minerals from the water, which are normally found in drinking water. For long-term use, consider adding a pinch of salt to every liter of distilled water to compensate for lost minerals.

Regularly inspect your distillation setup for any signs of wear or damage, especially if you're using it frequently. Always handle the system with care, especially when dealing with hot materials and steam.

By following these steps, you can create a reliable distillation system with everyday items, ensuring access to clean water in situations where traditional purification methods might not be available. While distillation is more time-consuming than other methods, its effectiveness against a broad range of impurities makes it invaluable in emergencies where water quality is compromised.

In situations where conventional water sources and purification methods are unavailable, natural materials such as sand, charcoal, and gravel can be lifesavers. These elements, readily found in nature, can be utilized to construct an improvised yet effective water filtration system.

Selecting the Right Materials:

- Sand: This fine material acts as a secondary filter, trapping smaller particles that bypass the initial gravel layers.
- Charcoal: Preferably use activated charcoal, known for its superior filtration properties, including removing toxins and improving water's taste. Natural charcoal, made from burnt wood, also works by absorbing harmful contaminants.
- Gravel: This is your preliminary filter layer, capturing larger debris and sediment. Utilize a variety of gravel sizes to enhance the filtering process, layering from coarser pieces on top to finer ones at the bottom, directly above the sand.

Steps to Assemble Your Filtration Device:

1. Container Preparation: Choose an appropriate container based on the amount of water you need to filter. This could range from a small plastic bottle for personal use to a larger bucket

for group needs. Modify the container by making small holes at the bottom to allow filtered water to exit. If using a bottle, invert it after cutting off the bottom to create a funnel shape.
2. Layering Materials: Construct your filter starting with coarse gravel at the new top, then a layer of finer gravel, followed by clean sand, ensuring each stratum is a few inches thick for adequate filtration. Cap off with a layer of crushed charcoal; while it should be fine, avoid powdering it to prevent water blockage.
3. Finalizing with Cloth: Cover the charcoal with a piece of clean fabric or a coffee filter. This prevents charcoal from mixing into the water and ensures an even spread across the filtration materials.
4. Operation and Maintenance: Gently pour water to be filtered through the setup, allowing it to trickle out the bottom slowly. The initial water might still be murky, serving to prime the filter — discard this. After initial use, maintain your filter by regularly replacing the charcoal and cleaning other materials to ensure effectiveness.

Although this homemade system significantly reduces many types of contamination, it might not remove all pathogens. To maximize safety, filtered water should subsequently be boiled or chemically treated, particularly if sourced from areas with high contamination risks. Always evaluate the clarity and taste of the filtered water, and when in doubt, treat it further to ensure its safety.

In constructing this natural filtration system, you're not just applying practical survival knowledge but also engaging with the environment thoughtfully. This method showcases the ingenuity required in survival situations and underscores the importance of adapting to available resources.

When it's time to bunker down or sketch out your long-term stay strategy, remember: water test kits aren't just fancy add-ons; they're your eyes in the murky waters of survival. They go beyond the basic look and taste test, diving deep into the unseen, checking for sneaky pathogens, chemicals, and metals. These kits might not be your first grab in a dash scenario, but they're pivotal for your ongoing safety checklist. Having them means you're not just surviving; you're thriving with knowledge on your water's condition, making informed calls on further purification needs.

Post-purification storage is another cornerstone. Don't slack here; proper storage means stopping contaminants at the door. Opt for sanitary, purpose-specific containers. Seal them as if your life depends on it (because, well, it does). Tag each with the purification date. Keep a strict timeline; a six-month refresh cycle ensures your water isn't just sitting there; it's waiting in prime condition.

Becoming a water safety guru isn't overnight work. It's a craft, honed over each camping stint, each home drill. It's about making purification second nature, so when the world tilts, your water smarts stay upright.

Sharing this wisdom? That's gold. Elevate your circle by turning knowledge into communal strength. Picture this: your backyard, a bunch of eager learners, and you, the water wizard, dispelling myths, spreading skills. It's not just about safety; it's about empowerment, building a network ready to face whatever the skies throw down.

But don't hit pause on learning. The water purification realm is ever-evolving — new gadgets, new methods, new threats. Stay in student mode, always a step ahead, ensuring your techniques are top-tier, your gear state-of-the-art.

Adaptability in water purification isn't just handy; it's crucial. Conditions change, supplies vary, emergencies arise. Your ability to switch tactics, to use what's at hand smartly, defines your survival. It's about making the right call for the moment, ensuring your water strategy is as fluid as the resource itself.

Understanding water purification and storage is non-negotiable, the very essence of survival, especially when off-grid living isn't a choice but a necessity. It's not about hoarding but about strategic, safe stocking. Your selection of containers is critical – aim for non-toxic, sturdy materials that promise purity and longevity. These are the guardians of your water's safety.

Before you even eye those storage barrels, get your source water right. Every source, be it a whispering stream, a rooftop channel, or a tap, must pass the gauntlet of purification. Your arsenal? Boiling, chemicals, filters. Then, with urgency and precision, transition this purified treasure into your chosen sanctuaries, safeguarding it from the insidious creep of contaminants.

This isn't about compiling a shopping list of gear or memorizing a survival manual. It's about ingraining a lifestyle, embedding a mindset of vigilance, preparedness, and adaptability. It's about not just enduring, but thriving, knowing that come what may, your water — the essence of life — is as resilient as your spirit.

Where you store your water can be as pivotal as how you store it. Aim for a space that's consistently cool and shaded, away from direct sunlight or fluctuating temperatures, which can promote algae growth and degrade water quality over time. Think: a dark corner of a basement, an insulated closet, or even an underground space if available.

Even with the best storage practices, water doesn't last forever. Implement a system to rotate your supply every six to twelve months. Use older stores for non-consumption purposes and replenish them with freshly purified water. This cycle ensures your supply remains fresh and potable.

Don't just stash your water away and forget about it. Label each container with the date of storage and the method of purification. This simple step helps track the age and safety of your water, making rotation schedules easier to manage.

Protecting your water from new contaminants post-storage is vital. Regularly inspect your containers for integrity, ensuring lids are sealed tight and there's no sign of leakage or damage. Store your water away from any substances that could pose a risk of contamination, such as gasoline, pesticides, or cleaning supplies.

In scenarios requiring ultra-long-term storage, you might consider adding water stabilizers. These can extend the life of your stored water, but ensure they're safe and intended for consumption. Remember, though, no additive is a substitute for regular water rotation and testing.

Speaking of testing, make water quality checks a routine part of your schedule. Use water testing kits to screen for bacteria, chemicals, and other pollutants. If you detect any contamination, treat the water again before use, following the proper purification steps. Here's how you can ensure your stored water remains safe for consumption:

Using a Water Testing Kit.

1. Gather Your Supplies: Obtain a water testing kit suitable for detecting bacteria, chemicals, and other common pollutants. These kits are available online or at local home improvement stores.
2. Collect a Water Sample: Choose a sample from your stored water supply. If you're testing water from a large container, use a clean, sterilized cup or ladle to collect the water. Avoid touching the inside of the container or the water with your hands to prevent contamination.
3. Conduct the Test: Follow the instructions provided with your testing kit. This typically involves adding a few drops of testing solution or dipping a test strip into your water sample. Ensure you follow the timing guidelines closely for accurate results.
4. Interpret the Results: Compare the color change of the test strip or solution to the color chart included in your kit. This will indicate the presence and level of contaminants such as bacteria, lead, pesticides, or nitrates.
5. Act Based on Results: If the test indicates contamination:
6. Boil the water for at least one minute (or three minutes at higher altitudes) to kill pathogens.
7. If boiling is not possible, use a new batch of chemical purification tablets following the package instructions.
8. Re-filter the water using a reliable filtration system if available.
9. Document Your Findings: Keep a record of the test results, including the date and the condition of the water. This helps track the quality of your water over time and identify when it's time to rotate your supply.
10. Repeat Regularly: Establish a routine for testing different parts of your water supply. Frequent checks ensure ongoing safety and alert you to potential issues before they become serious threats.

By following these steps, you maintain a clear picture of your water's safety and can take immediate action if any issues arise, ensuring your survival supply remains clean and drinkable.

Your storage methods might need to adapt depending on your local climate. In colder regions, avoid filling containers to the brim to prevent cracking from expansion if the water freezes. In hotter areas, prioritize locations that remain cool to discourage evaporation and maintain water integrity.

Diversify your storage locations within your home or property to ensure access in various scenarios. If one area becomes compromised due to flooding, structural damage, or other unforeseen circumstances, having backups can be life-saving.

For those pressed for space, get creative with your storage solutions. Water bladders designed to slide under beds or stackable water bricks can significantly increase your storage capacity without overtaking living areas.

If living in a community or shared space, coordinate with others on storage strategies. Collective efforts in purifying, storing, and rotating water supplies not only lighten the workload but also enhance the overall resilience of your community.

By following these guidelines, integrating regular practice and updates into your routine, and adapting to specific needs and environments, you ensure continuous access to this vital resource, fortifying your independence and readiness in facing any situation head-on.

Hit the road with confidence by stashing durable water containers in your ride. Think ahead, rotate that stash to keep your water supply fresh. This isn't just a tip, it's a must-do for anyone serious about their bug-out strategy.

Don't just sit back and hope for the best with your water. Have a plan B, C, and D for purifying it. Tainted supply? No sweat. Whip out those purification tablets, fire up a portable stove for boiling, or use that filter you smartly packed. Your safety's non-negotiable, and so is clean drinking water.

Wrapping this up, remember: managing your water isn't a once-and-done deal; it's an ongoing mission. The skills and strategies laid out here are your tools for survival – not just fancy words on a page. Arm yourself with knowledge, keep your supplies in check, and stay ready. Water's more than a necessity; it's your lifeline in times of crisis. Keep drilling these practices until they're second nature, and ensure your family does the same. Stay legal, stay smart, and above all, stay safe. If you've absorbed even half of what's in this guide, you're already ahead of the curve. Keep this manual close, but your wits closer. In the end, it's not just about making it through; it's about thriving when others are merely surviving.

Book 3: First Aid - More Than a Kit, A Lifeline

Welcome to your no-frills guide to survival first aid kits. Our ur world is unpredictable, filled with unforeseen challenges, from natural disasters to sudden accidents. This book is crafted for individuals who understand that being prepared is not optional—it's essential. Imagine finding yourself in a remote location, far from help, when trouble strikes. In that moment, your first aid kit is more than just supplies; it's your ticket to making it through.

We're cutting straight to the chase here. You, the reader, are the kind of person who doesn't wait for others to fix problems. Whether you're a long-time prepper or just awakening to the importance of preparedness, this guide is for you. It's about practicality, empowerment, and real-world applications. We'll dive into what makes a first aid kit not just good, but indispensable, and how to use it effectively in a crisis. This isn't about hypotheticals; it's about readying yourself for the real challenges that life can throw your way, ensuring that when faced with adversity, you're not just reacting, you're responding with confidence and competence.

When you're piecing together a survival first aid kit, you're not just packing; you're planning for the unplanned. This kit isn't about filling a box with medical supplies; it's your backup, your plan B, when you're miles from anywhere.

First up, let's talk essentials: bandages, antiseptics, tools. Your kit should have a variety of bandages – not just sizes, but types. Fabric ones are durable, waterproof ones keep moisture out, and gauze pads and rolls come in handy for larger wounds. Think about what you might face and pack accordingly.

Now, wounds need cleaning, so antiseptics are a must. Alcohol wipes kill germs on the skin, iodine can disinfect cuts, and hand sanitizer is your go-to when water's scarce. Here's a pro tip: those little single-use packets? Gold. They save space and one is usually enough for one clean-up job.

Closure devices like butterfly closures can pinch a cut closed, medical tape holds dressings in place, and believe it or not, super glue can seal wounds. But use the glue wisely – it's not for deep or infected wounds.

Tools – tweezers for splinters, scissors for cutting cloth or bandages, safety pins for... well, a lot of things, and a needle – not just for sutures (which you should only attempt if you're trained) but also for gear repairs.

Protection is non-negotiable. Gloves keep both you and the injured party safe from infection. Throw in some surgical masks for respiratory protection – they're light and don't take up much space.

Medications – cover the basics: pain relief (like ibuprofen), something for allergies (antihistamines), and your own prescriptions. Here's something actionable: list your meds on a small card and keep it in the kit. That way, if you're incapacitated, someone else will know what you need.

Specialized gear depends on where you are. A tourniquet for heavy bleeding can save a life, but only if you know how to use it correctly. Splints for breaks, thermal blankets for shock – these are essentials if you're heading into rough territory.

Let's talk about the extras that could make a difference. Navigation is more than just not getting lost; it's about making smart choices. A compass and waterproof maps of your area are basics. But don't just pack them; know how to use them. Practice before you head out.

Signal mirrors and whistles are your "hey over here" tools. Three blasts on a whistle is a universal distress signal and much easier on the throat than shouting. A signal mirror can catch light and attention from miles away – learn the S.O.S signal.

Customize your kit for the trip and for yourself. Allergic to bee stings? Pack an EpiPen. Asthma? Make sure your inhaler's in there. Adjust your kit based on where you're going and what you'll be doing.

Storage – keep your kit in a sturdy, waterproof container, and make sure everyone knows where it is. Regular checks every six months ensure everything's in date and in working order.

Now, having a kit is one thing, but knowing how to use it? That's where the real prep happens. Take a first aid course, then take it again. Practice makes prepared. Set up scenarios and run through them until everyone's comfortable with what to do.

Nutrition might not seem like a first aid issue, but in a survival situation, it is. High-energy snacks, a solid multivitamin, and electrolyte packets can be lifesavers. Your body's your best survival tool – keep it fueled.

Water – you need it, so plan for it. Pack water purification tablets and know where you can find water in your area. Dehydration isn't just about being thirsty; it can cloud your judgment and slow you down.

Now, let's talk size. Your kit should be big enough to hold what you need but small enough to be manageable. Every item should earn its place. Ask yourself: Do I need this? Will I know how to use it? Is there something else that could do the job better?

Finally, your first aid kit is more than supplies; it's a mindset. Packing this kit means you're thinking ahead, planning for the just-in-case, and ready to take care of yourself and others. It's about being prepared, sure, but it's also about being proactive. Don't just pack it; know it, inside and out.

Here's the bottom line: your survival first aid kit is your lifeline out there in the wild, the difference between a minor hiccup and a major ordeal. Pack it with care, with knowledge, and with intention.

Then, hopefully, you

Here's a concise list of essentials for your first aid kit:

- Bandages: Assorted sizes of adhesive bandages, gauze pads, and rolls.
- Closure Devices: Butterfly closures, medical tape, and super glue.
- Antiseptics: Alcohol wipes, iodine solution, and hand sanitizer.
- Tools: Tweezers, scissors, safety pins, and a sewing needle.
- Protection: Disposable gloves and face masks.
- Medications: Pain relievers, antihistamines, and personal prescription medications.
- Specialized Equipment: Tourniquets, splints, and thermal blankets.
- Instructional Materials: Waterproof first aid guidebook or instruction cards.
- Navigation Tools: Compass and waterproof local area maps.
- Signaling Devices: Signal mirror and a loud whistle.
- Water Purification: Tablets or drops.
- Nutritional Supplements: High-calorie snacks, multivitamins, and electrolyte packets.

Your survival first aid kit isn't just a set of tools and supplies; think of it as a dynamic companion that evolves with you. Every trip into the great outdoors, every unforeseen situation, and every bit of new knowledge gained should inform the contents of your kit. It's not about having a one-size-fits-all solution but developing a personalized arsenal that reflects real-world experience and learning. When you use an item from your kit, take it as a learning opportunity. Was it effective? Was anything missing? Each experience is a chance to refine your kit, making it more effective for the next time.

As you grow, so should your kit. New skills might mean new tools. Maybe you've taken a wilderness medicine course and can now include more advanced medical supplies. Perhaps a close call has shown you the need for specific items you hadn't considered before. Changes in your physical condition could also necessitate updates; for example, new allergies require new medications. Keep this evolution ongoing; complacency has no place in preparedness.

Let's not forget the world around us doesn't stay static. Environmental shifts, whether due to the changing seasons or moving to a new area, can significantly impact the utility of your kit. A kit perfect for the humid, bug-ridden summers might not serve as well in the freezing winters. Similarly, what works in a rural setting may fall short in urban emergencies. Update your kit to match your surroundings and the specific challenges they present.

Sharing knowledge is just as crucial as acquiring it. Your journey towards self-reliance and preparedness shouldn't be a solitary trek. Engaging with your community, sharing your experiences, and learning from others not only strengthens your own skills but also enhances the collective resilience. Teach what you know, whether it's bandaging a wound, purifying water, or signaling for help. Every piece of knowledge shared is a step towards a more prepared and resilient community.

Now, onto your first aid kit and beyond—this isn't just any kit. It's a clear signal of your readiness to tackle unexpected challenges head-on. It's an embodiment of your commitment to safety and well-being, not just for yourself but for those around you. Treat this kit with the respect it deserves; maintain it, know it inside out, and it will be there for you when the chips are down.

Heading into specifics, understanding first aid basics is non-negotiable. Immediate care, whether for minor injuries or life-threatening conditions, can significantly impact outcomes. First aid is about preserving life, preventing further harm, and fostering recovery. Grasping these basics could be what stands between a manageable situation and a dire one.

Safety first, always. Before you dive into aid, assess your surroundings. Ensure that by helping, you're not putting yourself at unnecessary risk. Check for immediate dangers like fire, hazardous materials, or signs of structural instability. This step is crucial—there's no help to be offered if you end up in harm's way yourself.

When it comes to assessing safety, here's what to consider:

- Immediate Dangers: Before you do anything, stop and look around. Are there risks like fire, falling debris, or dangerous animals?
- Environmental Conditions: Be aware of the weather and terrain. Extreme heat or cold can impact both you and the victim.
- Bystanders: Use them as resources. What do they know? Can they help?
- Continuous Assessment: Keep evaluating the situation. Things can change quickly.

After ensuring safety, if emergency help is needed and reachable, make the call. Provide clear, concise information about your location and the nature of the emergencies. In remote settings, knowing how to signal effectively could be crucial.

Your primary assessment follows a critical sequence, often remembered as the ABCs: Airway, Breathing, Circulation. Is the airway clear? Is the person breathing? Do they have a pulse? These are your immediate checks.

If you encounter bleeding, control is key. Start with direct pressure and elevate the injury if possible. If the bleeding doesn't stop, you may need to consider additional measures like pressure points or, as a last resort, a tourniquet. But remember, certain actions, like applying a tourniquet, require proper knowledge and training to ensure they do more good than harm.

Here are the steps to control severe bleeding effectively:

11. Direct Pressure: Firmly apply pressure with a clean cloth or bandage.
12. Elevation: Raise the injury above the heart if possible.
13. Pressure Points: Use only if necessary and you know how.
14. Tourniquet: A final measure, used only when bleeding is life-threatening and other methods

haven't worked.

Understanding and practicing these steps can mean the difference when faced with a life-threatening situation. Regular refreshers in first aid principles and techniques can significantly boost your confidence and efficiency in emergency scenarios.

Treating Shock Effectively.

Shock is not merely a reaction; it's a critical, life-threatening condition, often the result of trauma, significant blood loss, or overwhelming stress. It leads to decreased blood flow and a lack of oxygen to vital organs, which can be fatal without timely intervention. When dealing with shock, every second counts. Your actions could be the lifeline the victim needs while professional medical help is on the way.

Here's an in-depth guide on managing shock effectively:

Clear Airway: Ensuring the victim can breathe is your top priority. If they're unconscious yet breathing, position them in the recovery position. This stance helps maintain a clear airway and prevents any blockage, which could be life-threatening.

Calm Presence: The power of your voice and demeanor cannot be overstated. A calm, reassuring presence can help stabilize the victim's heart rate and anxiety, aiding in shock management.

Conserve Heat: Shock can rapidly decrease body temperature, leading to hypothermia. Use blankets, coats, or even your body warmth to keep the victim warm and prevent their condition from worsening.

Fluid Restriction: Although they may request water, refrain from giving any fluids by mouth. There's a significant risk they could choke or aspirate, especially if they lose consciousness.

Vital Monitoring: If you're trained, monitor their vital signs. This can provide essential insights into their condition and help gauge the severity of the shock.

Leg Elevation: This can help blood return to the heart and improve circulation to the brain and vital organs. However, avoid this if you suspect spinal injuries or broken bones.

Vomiting Precautions: Shock can induce nausea. If the victim shows signs of vomiting, especially if semi-conscious or unconscious, turn their head to the side to ensure the airway remains clear.

Constant Vigilance: Keep a watchful eye on the victim for any changes. Conditions can deteriorate swiftly; continuous observation helps you respond promptly to any new signs of distress.

Emergency Services: Contacting emergency medical services should be done immediately. Provide them with comprehensive information about the situation and what you have observed and done so far.

In addition to shock management, dealing with physical injuries like fractures and sprains correctly is paramount. Immobilizing the affected area is crucial in cases of suspected fractures. Utilize available resources – from sticks to rolled-up magazines – to create stabilizing splints. For sprains, the RICE method remains a cornerstone of initial treatment:

- Rest: It's essential for recovery. Keep the injured part still and supported.
- Ice: Applying ice helps reduce swelling and dulls pain. However, never place ice directly on the skin.
- Compression: Use bandages to apply gentle pressure. This helps keep swelling in check.
- Elevation: Raising the injury above heart level aids in reducing swelling by improving blood flow away from the injured area.

The correct application of the RICE method can substantially mitigate the pain and swelling of minor injuries and is a pivotal part of the healing process.

Addressing burns promptly and effectively is also crucial. For minor burns, immediate cooling under running water significantly reduces injury severity and alleviates pain. Remember, cooling the burn helps halt the damage progression. It's vital, however, to ensure that the water is cool, not cold, to avoid causing additional harm.

Following cooling, it's important to cover the burn with a sterile, non-stick bandage or clean cloth. This layer acts as a barrier against infection and reduces pain by shielding the burned area from air exposure.

Monitoring the burn for signs of infection, such as increased redness, swelling, or the presence of pus, is critical. Infections can complicate the healing process and necessitate medical intervention.

For pain management, over-the-counter medications can be effective but ensure they do not clash with the victim's medical history or current medications.

Finally, as the burn begins to heal, maintain the skin's moisture with appropriate creams or lotions and protect the area from sunlight to prevent scarring and promote better healing.

In all instances, whether managing shock, treating fractures and sprains, or caring for burns, your knowledge, preparedness, and calm demeanor play vital roles in the outcome. By adhering to these detailed steps, you equip yourself with the tools and understanding necessary to provide effective first aid. This not only aids in the immediate care of the victim but also bridges the crucial gap until professional medical help is available. Remember, in the realm of emergency preparedness and response, your actions can significantly alter the course of recovery, demonstrating the profound impact of well-informed, decisive action in critical moments.

CPR.

CPR, or Cardiopulmonary Resuscitation, becomes vital when someone's heartbeat or breathing has halted, usually from a cardiac arrest. Here's how to step in effectively:

Assess the Situation: Make sure you're not putting yourself in harm's way. Then, see if the person responds to your shouts or gentle shoulder taps. If there's no reaction, they need your help.

Alert Authorities: Time is against you. If the person is unresponsive, get emergency services on the line ASAP. If you're solo, give CPR a solid two minutes before you call for help.

Prep the Person: They need to be lying on their back on something firm. You'll be doing chest compressions, and a soft surface won't do.

Start Compressions: Position your hands one on top of the other, mid-chest. Lock your elbows and push down hard and fast, aiming for 100 to 120 compressions per minute. Depth is crucial – at least two inches. Let the chest rise fully between pushes.

Give Rescue Breaths: If you've been trained, after thirty compressions, tilt their head back, lift their chin, and deliver two breaths into their mouth, ensuring no air escapes. If you're not comfortable or trained, stick to chest compressions.

Keep Going: Swap between thirty compressions and two breaths. Don't stop unless they start to show signs of life, or help takes over.

Use an AED: If there's an Automated External Defibrillator handy, use it. Follow the spoken prompts it will give you.

Infection prevention post-injury is just as crucial as initial treatment. Here's how to keep wounds clean and reduce infection risk:

Clean Hands: Before you touch the wound or anything related, wash your hands. This is non-negotiable – cleanliness is your first line of defense against infection.

Prepare the Wound: Clear away debris or dirt with clean water. Stay away from hydrogen peroxide or iodine; they're harsh and can irritate the wound further. Simple running water will do the trick for minor cuts.

Disinfect Around the Wound: Use antiseptic solutions or wipes on the skin around the injury. Direct application of harsh substances like alcohol can damage tissue and worsen pain, so avoid that.

Control Bleeding: Apply mild pressure with a clean cloth or sterile dressing to stem any bleeding. Keep at it until it stops.

Bandage Properly: Once you've cleaned and dried the area, cover it with a sterile dressing. This keeps out bacteria and helps the healing process. Remember to change the dressing in case it gets wet or dirty.

Watch for Trouble: Be vigilant for signs of infection such as increased redness, swelling, or pus.

These are red flags.

Dry and Clean: Keep the wound dry and clean as it heals. No baths or swimming until it's fully healed.

Recognizing signs of serious conditions like heart attacks, strokes, and severe allergic reactions is life-saving:

Heart Attack: Look for chest pain, discomfort in other upper body areas, breathlessness, cold sweats, nausea, or dizziness. Act immediately – call for help. If it's safe, and the person is conscious, an aspirin might help reduce clotting, but only if there are no contraindications.

Stroke: Remember FAST: Facial drooping, Arm or leg weakness and Speech difficulty… Time to call for help. Time is brain; the faster you act, the better their recovery chances.

Diabetic Emergencies: For low blood sugar, if they're awake, get them something sweet. For high sugar, they might need insulin, but that's usually for them or a professional to administer.

Anaphylaxis: Hard breathing, swollen face or throat, rapid heartbeat, or dizziness signal this severe allergic reaction. If they have an EpiPen and know they're allergic, help them use it, then call for emergency assistance.

Understanding and acting upon these protocols can dramatically increase survival rates and recovery quality. These skills are critical, not only for those living off-grid but for anyone who values preparedness and self-reliance in today's unpredictable world. By mastering these techniques, you're not just ready to protect yourself; you're equipped to safeguard those around you, turning potentially dire situations into manageable ones.

Let's ensure that your emergency preparedness and off-grid communication strategies are not just plans but practiced, ready responses for when the unexpected becomes today's reality. Remember, in emergencies, knowledge is as crucial as action. Your readiness can be the difference between life and despair, between recovery and catastrophe. So, take this knowledge, hone it, and be prepared to use it.

When confronting environmental injuries, swift action is key. Here's what you need to know for conditions caused by extreme temperatures.

If someone's hit with heatstroke, a severe condition where the body overheats, they'll show signs like skyrocketing body temperature, confusion, and a fast heartbeat. Here's what to do: move them to a shady or air-conditioned place immediately. Peel off any heavy clothing and cool them down with damp cloths or a cold bath. If they're confused or unconscious, don't give them anything to drink. Time is critical — call for emergency help right away.

Facing heat exhaustion, you'll see intense sweating, weakness, and skin that's cold and pale. The action plan is straightforward: get them into a cool area, have them lie down, loosen clothing, and apply cool, wet cloths. If they're conscious, let them sip water slowly. Keep an eye on them; if there's no improvement or they start vomiting, seek medical attention without delay.

In the cold, hypothermia sneaks up when the body loses heat faster than it can generate it, leading to dangerously low body temperature. Watch for uncontrollable shivering, tiredness, and confusion. The priority is to warm them up — move them indoors or to a sheltered area. Focus on warming the core of their body first (think chest, neck, head, and groin). If possible, use an electric blanket or body heat. If they're awake, offer warm, non-alcoholic drinks. Hypothermia is serious — get medical help immediately.

Frostbite is the freezing of body tissue and looks like numbness and skin turning waxy or a strange color. First, get them somewhere warm. Then, gently warm the affected area in warm (not hot) water, or with body heat — never rub or massage frostbitten skin. Seek medical care urgently if frostbite is severe, indicated by blistering or changes in skin color.

Creating an emergency medical plan isn't just good sense — it's a necessity. Know the fastest route to the nearest hospital, have a list of crucial emergency contacts ready, and ensure that anyone you're with knows about any medical conditions that might need sudden attention.

Training is everything. Staying sharp with regular, certified first aid courses means you're always ready to respond. And it's not just for your sake — teaching your skills to family and friends extends that preparedness to your wider circle, making everyone safer.

But remember, knowledge alone isn't enough. Practice your first aid skills regularly to build confidence and ensure you can act effectively in real-life situations. By consistently revising what you've learned and applying it, even when there's no emergency, you solidify those life-saving skills.

Spreading first aid knowledge is about more than just making yourself safer; it's about lifting the safety of everyone around you. Encourage others to learn and practice first aid. The more people know, the safer your community becomes.

In emergencies, stress can be as dangerous as any physical injury. Recogniing when you or someone else is stressed — signs like irritability, fatigue, or trouble concentrating — is the first step in tackling it. Just as you gear up physically, mental preparation is crucial. Understand the kinds of stress you might face and plan how to handle them.

Breathing can be a powerful tool against stress. The 4-7-8 technique — breathe in deeply for four seconds, hold for seven, and exhale slowly for eight — is simple yet effective. It's a quick way to calm down and refocus, essential in an emergency when clear thinking is critical.

By adopting these approaches, from managing environmental injuries to maintaining calm under pressure, you'll not only be safeguarding yourself but also those around you. Preparedness is not just a personal choice; it's a communal responsibility. In a world where the unexpected lies around every corner, being ready is the best defense.

In crisis situations, establishing a routine can anchor you, providing psychological stability and a semblance of control in the midst of chaos. Start your days with clear plans, even for basics like meals or rest. Having a structured plan helps map out your day, providing clarity and a sense of purpose.

Designate specific areas for different activities. This separation between sleeping, eating, and other tasks is more than just physical—it helps organize your thoughts and priorities. Tackle essential needs first: shelter, water, food. Knocking these off your list not only ensures survival but boosts your morale.

Don't neglect physical activity. Integrate simple exercises into your day to break monotony and maintain health. Include downtime activities like reading or writing to unwind and maintain mental health. Consistent sleep schedules reinforce your internal clock, offering comfort in the unpredictable.

Physical activity is a known stress-buster. Opt for exercises that are feasible—like walking, stretching, or body-weight routines. Keep sessions short but consistent to lift your spirits. Pick activities you enjoy; the pleasure enhances the stress relief. Whenever possible, exercise outside to compound the calming effects.

Don't let isolation amplify stress. Stay connected with your group or reach out via journals or messages if alone. Future sections of this guide will delve into communication techniques to bridge gaps between isolated individuals.

Zero in on factors within your control. List what you can manage—your actions, plans, outlook—and set achievable goals. Redirect from the uncontrollable to actionable steps, and don't hesitate to lean on your support network for help and brainstorming.

Mindfulness and meditation can be your sanctuary. Find a quiet spot, focus on your breath, and gently redirect when your mind wanders. Visualize serene landscapes to deepen relaxation. Start with short sessions, expanding as you grow more comfortable. This practice can be a cornerstone for stress management, grounding you in the present moment.

Lastly, never underestimate the role of nutrition and hydration in mental well-being. Balanced diets and regular water intake keep your body primed to handle stress better. Avoid excessive caffeine and sugar, which can heighten anxiety. Practice mindful eating to enhance your connection to food and to avoid stress-induced overeating.

This approach, blending routine, physical activity, connectivity, control, mindfulness, and nutrition, equips you to face emergency scenarios not just with physical preparedness but with mental resilience. By integrating these strategies into your daily life, you ensure that when emergencies arise, you're not just surviving; you're thriving, maintaining mental clarity and emotional balance even when the world seems to be crumbling around you.

Securing proper rest and a solid sleep routine is not just beneficial—it's critical, particularly when under the strain of emergencies. Establishing a sleep schedule where you hit the sack and wake up at the same hours daily, including weekends, can significantly improve your sleep quality and mental sharpness.

Transform your sleeping space into a calm sanctuary. Keep it cool, quiet, and dark to promote restful sleep. If noise and light intrude, consider using earplugs and blackout curtains or eye masks. Cut down on blue light exposure from screens before bed, as this can interfere with your sleep cycle. Introduce calming activities before bed like deep breathing, meditation, or some light stretching. This isn't just fluff; it's about telling your body it's time to power down. Be mindful of what you consume before bedtime. Avoid heavy meals, caffeine, and alcohol as they can disrupt your sleep, turning rest into a restless chore rather than a restorative necessity.

In survival situations, admitting you need help is not a sign of weakness but of practical intelligence. Whether signaling for rescue or leaning on your survival group for emotional support, knowing when and how to reach out is key. Post-crisis, it's equally important to address any lingering mental health concerns with a professional.

Treat each challenge as a lesson. Post-situation reflection is crucial. Ask yourself: What stress relief tactics worked? What didn't? Use these insights to refine your approach, bolstering your readiness for next time.

Small victories should be recognized and celebrated. They're not just feel-good moments; they're milestones of progress, essential for maintaining morale and momentum in the face of ongoing challenges. This is about recognizing the strength and perseverance it takes to move forward, one step at a time.

Understanding Post-Traumatic Growth (PTG) shifts the narrative from merely surviving to thriving after adversity. It's about seeing beyond the hardship to the personal growth that comes from overcoming tough situations. Reflect on past difficulties and identify how they've made you stronger or changed your outlook for the better. This isn't about glossing over pain but about finding a path to resilience and deeper understanding through it.

Embrace change and view challenges as stepping stones, not stumbling blocks. Open dialogue with supportive friends or family, or seek a professional's insight to help uncover the silver linings and growth opportunities in tough times. Make time for self-reflection. Growth is a journey, often without a clear end point, but every step forward counts.

This guide isn't just a set of instructions; it's a blueprint for building resilience and mental fortitude, ensuring that in the face of adversity, you're not just getting by—you're getting stronger. By adopting these strategies, you're preparing not just for the physical aspects of emergencies but for the mental and emotional challenges as well.

Staying physically active, even in the thick of stress, acts as a natural stress relief valve. Choose exercises that fit the situation—simple, equipment-free movements like walking, stretching, or bodyweight exercises. Consistency is more crucial than intensity; even short bouts of physical activity can elevate your mood and ease stress. Opt for activities you enjoy, making exercise a welcome break rather than a chore. Whenever possible, take your physical activity outdoors. The natural setting amplifies the relaxing effect, boosting your mental well-being.

Isolation can magnify stress. Maintain connections as much as possible. If you're physically alone, use journals or voice recordings as a means of expression and reflection. These can be vital for maintaining a sense of connectedness and for processing experiences.

Direct your focus toward what you can control: your preparation, your actions, and your attitude. Break overwhelming challenges into smaller, manageable tasks and celebrate the completion of each. This strategy not only keeps you grounded but also propels you forward, providing clear evidence of your capability and progress.

Incorporate mindfulness and meditation into your routine. Find a quiet spot, focus on your breathing, and allow this practice to anchor you in the present, offering a respite from the chaos. Starting with just a few minutes a day can make a significant difference in your stress levels and overall mental clarity.

Don't underestimate the impact of diet and hydration on your mental state. A balanced diet fuels the body and mind, keeping you at your best. Regular hydration keeps the mind clear and focused. Limiting stimulants like caffeine and sugar can also stave off spikes in anxiety and restlessness.

By weaving these practices into your daily life, you're not only gearing up to handle emergencies but also enhancing your day-to-day resilience and well-being. It's about creating a lifestyle that supports preparedness, resilience, and personal growth, ensuring you're ready for whatever lies ahead.

In wrapping up this guide, let's hammer home the essence of what it means to be truly prepared. This isn't just a book; it's a lifeline—one that you've taken the initiative to grasp firmly in your hands. Remember, the path to self-reliance in emergencies is paved with persistent practice, constant learning, and the willingness to adapt. Don't let this be the last time you open these pages. Make it a regular exercise to review what you've learned and put it into practice. Preparedness is not a one-off task but a continuous commitment to your and your loved ones' safety and well-being.

Moving forward, take the knowledge you've absorbed and step up as a leader in your community.

Your newfound skills are not just for your benefit but can serve as a beacon of strength and resilience for those around you. Teach, share, and encourage others to join you on this journey of preparedness. The more people who are prepared, the stronger your community will be when faced with adversity. Remember, in times of crisis, we're stronger together.

Finally, stay curious and proactive. The world of first aid and emergency preparedness is ever-evolving, and so should you be. Keep abreast of the latest techniques, refresh your supplies regularly, and reassess your emergency plans as your life and the world around you change. Your readiness to face unforeseen challenges head-on will not only bolster your confidence but could one day save lives. Carry forward the spirit of preparedness, and let it guide you through life's unexpected turns with strength and certainty.

Book 4: Your Castle: Defense and Shelter

The importance of this book and the concept of home preparedness cannot be overstated, especially in today's unpredictable world. Natural disasters, unforeseen emergencies, and security threats are becoming increasingly common, making it essential for individuals and families to be well-prepared and resilient. This book serves as a crucial guide, equipping you with the knowledge, strategies, and practical steps to transform your home into a safe haven. It's not just about reacting to emergencies; it's about proactive preparation, ensuring that you and your loved ones remain safe, secure, and self-reliant in various scenarios. By fostering a culture of preparedness, you can significantly reduce panic and helplessness, turning potential chaos into manageable situations.

Having your home prepared means more than just stocking up on emergency supplies; it's about creating a comprehensive plan that encompasses all aspects of safety and security. This book delves into the nuances of home fortification, from securing entry points and establishing a secure perimeter to integrating advanced surveillance systems and fostering a resilient mindset among household members. It's a holistic approach that covers not only physical defenses but also psychological readiness, ensuring that you are mentally and physically equipped to handle unexpected situations. The peace of mind that comes from knowing you have taken all possible measures to protect your home and family is invaluable.

Moreover, this book emphasizes the importance of community and collective safety. In times of crisis, having a coordinated approach with neighbors and local authorities can significantly enhance the effectiveness of your preparedness plan. It fosters a sense of unity and support, ensuring that you are not alone in facing emergencies. By applying the principles and techniques outlined in this guide, you can contribute to a safer, more resilient community. The empowerment that comes from being prepared not only benefits you and your family but also strengthens the entire neighborhood, creating a collaborative environment ready to face and overcome any challenge.

In the landscape of home security, navigating the maze of legal stipulations is a must. Whether it's defensive tools, surveillance tech, or tweaks to your home's architecture, ensure every measure complies with local ordinances. This diligence safeguards your sanctuary's legality, fortifying not just its walls but its legal standing as well.

Transform your home from a mere shelter to a bastion of safety. This transcends installing locks and cameras; it's about a comprehensive fortification strategy that envelops both the tangible – like reinforced doors – and the intangible, such as the resilience and readiness of those within. Such a thorough approach wards off potential intruders and instills a profound sense of security and readiness in your household.

Heating your space without fire is not just innovative; it's a survival skill. Especially under the constraints of an emergency, where traditional heating may be off-limits, understanding and applying alternative heating methods is vital. This ensures comfort and prevents the life-threatening risks associated with cold, such as hypothermia, underlining the importance of non-traditional warmth sources when conventional options are unfeasible.

Insulating your home effectively is the cornerstone of maintaining warmth without relying on fire. Here's how you can lock in the heat:

Window Insulation: Seal windows with bubble wrap or plastic film to combat heat loss.

Door Draft Stoppers: Employ draft stoppers to eliminate cold air seepage under doors.

Wall and Floor Insulation: Utilize thick curtains and heavy rugs to add insulation layers, trapping heat indoors.

Layering is a tried-and-tested method for personal warmth. Start with a base layer that pulls sweat away, add a middle layer like fleece to trap warm air, and finish with a protective outer layer to shield against wind and water.

Mylar blankets are a critical emergency asset, reflecting back body heat and significantly boosting warmth without the need for fire, making them perfect for crisis situations.

Harness the power of passive solar heating to warm your living space. Direct sunlight through south-facing windows can provide natural warmth without any cost. Clear any obstructions, optimize window treatments to capture as much sunlight as possible, and use reflective materials to enhance the light's reach and warmth within your home. At dusk, close curtains to retain the heat. Incorporating materials like stone or brick that absorb and slowly release heat can further stabilize indoor temperatures.

For those of you living off the grid or simply aiming for self-sufficiency in emergencies, understanding your heating options can make all the difference. Here's a deeper look into non-conventional heat sources and how to effectively employ them.

Starting with Chemical Heat Packs, these are not just gadgets; they are lifelines in frigid conditions. Typically filled with a solution like sodium acetate, they spring into action once the metal disc inside is flexed, initiating a heat-releasing crystallization process. Picture this: you're caught in an unexpected snowstorm; your fingers are numbing, risking frostbite. Activating a chemical heat pack can turn the tide, warming your hands, and possibly saving fingers from frost damage. Post-use, don't discard them. Boil them until the contents turn liquid again and cool down for reuse. This cycle of activation and reactivation can be a continuous source of heat in prolonged power outages.

Gel Heat Packs come next, primarily used for larger areas needing warmth or healing. Before you hunker down for a cold night in a blackout, heating a gel pack in the last minutes of microwave power could provide prolonged warmth. Imagine heating one, wrapping it in a towel, and placing it at your feet under the blankets. It can mean the difference between a sleepless night shivering and a rested morning. Always monitor the heating time to avoid explosions or leaks, and remember, safety first: never place a super-heated pack directly against the skin.

Now, onto Grain-Filled Heat Packs, which are essentially nature's answer to sustained warmth. They can be homemade, filled with rice, wheat, or flaxseed - perfect for those who value natural, renewable resources. Here's a practical application: say a winter storm has knocked out power, and you need to keep a baby warm. A microwaved grain-filled pack, securely wrapped, can maintain a crib's warmth for hours. It's a sustainable way to harness heat, with the added benefit of customizing sizes and shapes to fit your needs, like a long, slim pack for wrapping around frozen pipes or a small, square one for pocket warmth.

When employing these heat sources, wrap them in cloth to avoid skin burns and ensure they're intact to prevent spillage. Regular checks for wear and tear are essential to maintain their safety and effectiveness.

Expanding on Body Heat Sharing, it's a method as old as humanity but crucial in survival situations. Imagine your car breaks down in a blizzard. The engine's dead, and so is the heater. Grouping together, using emergency blankets, and minimizing space between individuals conserves heat exponentially compared to when everyone is isolated. Employ sleeping bags, layer clothing, and use emergency blankets to maximize this effect. It's this collective warmth that could prevent hypothermia until rescue arrives.

Lastly, consider Solar Air Heaters. They're not just eco-friendly; they are power outage-proof sources of warmth. Constructing one can be a rewarding DIY project. Imagine using recycled aluminum cans, painted black, housed in a wooden frame under a clear polycarbonate cover. Positioned to catch the winter sun, this setup can channel warm air into your home without a single watt of electricity. For those looking to cut energy costs or simply extend their independence from the grid, building a solar air heater could be your next weekend project.

By applying these insights and examples to your emergency preparedness strategy, you're not just planning; you're ensuring comfort, safety, and a degree of warmth that can make all the difference in challenging situations. Remember, the goal here is not just survival but maintaining a standard of living that keeps spirits high and bodies warm, even when the mercury drops. Through understanding and preparation, you turn potential crises into manageable situations, embodying the true spirit of self-reliance and resilience.

Integrating thermal mass into your home goes beyond typical construction; it taps into ancient methods updated for modern resilience. These materials—bricks, stones, and concrete—aren't just structural; they're part of a strategic approach to temperature control, acting as natural heat reservoirs.

Imagine your home as a sponge for solar energy. During the sun's peak hours, thermal mass materials absorb and store this warmth. Even after the sun sets, they continue to release this stored energy, maintaining a warm, stable indoor environment as outside temperatures plummet. The gradual heat release from these materials can make a noticeable difference in your comfort levels and energy bills.

For instance, consider the case of the Anderson family, who live in a temperate climate with cold winters. They renovated their 1970s home to include a large stone wall in their south-facing sunroom and added concrete flooring throughout the living areas. During winter, the sunroom's stone wall absorbs heat throughout the day. At night, when the family gathers in the sunroom, it's comfortably warm without any additional heating. This passive heating approach has cut their winter energy costs by 30%.

Here's how you can apply these principles in your own home:

Positioning for Optimal Sun Exposure: Place thermal mass materials like bricks or concrete slabs where they can receive uninterrupted sunlight, especially near south-facing windows.

Enhancing Heat Absorption: Consider painting these materials dark colors, which naturally absorb more heat.

Insulation Matters: Surround thermal mass with adequate insulation to prevent heat escape. This ensures the stored warmth is effectively released indoors rather than seeping out.

Implementing thermal mass is just part of a broader strategy of using the earth itself for insulation and temperature regulation, known as earth sheltering. This approach involves using the earth's steady subterranean temperature to keep your home's interior more constant, reducing the need for mechanical heating and cooling.

Consider the case of Hillside Haven, an exemplary model of an earth-sheltered abode nestled into a gentle Midwestern slope. The homeowners, a couple deeply committed to the ethos of sustainable and efficient living, deliberately selected this earth-sheltering architectural style for its notable energy efficiency and minimal ecological footprint. The design ingeniously utilizes the earth's natural insulation properties, resulting in significant energy savings and an impressive reduction in the home's environmental impact.

The strategic positioning of the home maximizes its southern exposure, allowing large, well-placed windows to soak up the winter sun, harnessing passive solar heating to its fullest potential. This not only floods the living spaces with natural light but also significantly reduces the need for artificial heating during the colder months. Conversely, the earth berms, hugging the north, east, and west flanks of the home, shield it from harsh winds and summer heat, maintaining comfortable indoor temperatures throughout the year. The clever design essentially eliminates the need for conventional air conditioning, even in the height of summer.

For individuals considering the path of earth sheltering, here are expanded guidelines:

Site Selection: The choice of location is paramount. Look for land with the right kind of soil, adequate drainage to prevent waterlogging, and optimal exposure to sunlight, especially if aiming for passive solar benefits. The site should complement the natural advantages of earth sheltering, reducing construction costs and maximizing environmental harmony.

Design and Orientation: Your home should be more than a place of residence; it should be a cog in the wheel of sustainability. When planning the design, focus on harnessing passive solar energy by positioning living spaces and windows to face southward. This not only provides free heating and lighting but also contributes to your home's overall energy efficiency.

Insulation and Waterproofing: These are the shields that protect your sanctuary from the elements. Invest in high-quality insulation and waterproofing to prevent moisture intrusion, which is particularly crucial in subterranean living spaces. Proper insulation keeps the warmth in during winter and out during summer, while effective waterproofing guards against dampness, ensuring the longevity and comfort of your home.

Ventilation: Good air quality and humidity control are essential for a healthy living environment, especially in an earth-sheltered home where airflow might be restricted. Design a ventilation system that maintains fresh air circulation and manages moisture, preventing condensation and mold growth.

Incorporating the principles of thermal mass and earth sheltering is not limited to new constructions alone. Existing buildings can be retrofitted to embrace these concepts. For instance, integrating thermal mass into an existing structure might involve laying down a new tile floor in areas that receive ample sunlight, thus storing heat during the day and releasing it at night. Alternatively, one could add a sunroom facing the south to serve as a passive solar heater during the colder months.

Similarly, earth sheltering techniques can be adapted to existing homes. Simple measures like berming earth against a north-facing wall can significantly reduce heat loss and protect against cold winds. These adaptations, while seemingly minor, can lead to substantial improvements in energy efficiency and living comfort.

By embracing the holistic approach of earth sheltering and thermal mass, exemplified by Hillside Haven, homeowners can achieve a living space that is not only in tune with nature but also self-regulating, comfortable, and profoundly sustainable. This case study serves as a blueprint for those aspiring to create a home that blends seamlessly with the environment, offering lessons in the art of living harmoniously with our planet while forging a path towards true self-sufficiency and resilience.

By embracing these methods, you align with a philosophy that views your home not as separate from nature but as a part of it. This approach not only provides a more comfortable living space but also reduces your environmental footprint, a testament to the blend of traditional wisdom and modern innovation that defines true self-reliance.

These examples and instructions aren't just hypothetical; they are real steps that real people—just like you—have taken to make their homes more sustainable, efficient, and aligned with the rhythms of the natural world. By integrating these practices into your life, you're not just preparing for emergencies; you're reshaping your daily living into a model of sustainable, self-reliant living.

Adopting thermal mass and earth sheltering principles aligns with a broader movement towards sustainability and energy independence. By understanding and applying these concepts, you're not only crafting a living space that stands up to the challenges of climate and energy scarcity, but you're also setting a standard for future generations. It's a testament to the philosophy that our homes should work with the environment, not against it. This approach extends beyond personal benefit, contributing to a larger ethos of conservation and responsible stewardship of our planet. Embracing these principles doesn't just change how we live; it transforms how we relate to the world around us, turning our homes into beacons of resilience and sustainability.

Building Windbreaks.

Building windbreaks around your property can dramatically shield your living space from the biting cold winds, which not only plummet the outdoor temperature but can seep into your home, escalating your heating needs and costs. Here's how you can establish both natural and constructed wind defenses effectively.

For natural protection, plant a series of dense trees and shrubs on the wind-facing side of your land. Opt for evergreens and thick, bushy trees for a year-round barrier.L If you're in a position to plan according to the landscape, utilize natural formations like hills which can naturally break the wind's force before it reaches your home.

If you prefer or need immediate results, building fences or walls can offer an instant windbreak. Materials can range based on what's locally available or what fits your budget—wood, concrete, or even upcycled materials. When adding outbuildings like garages or sheds, position them strategically to act as large-scale wind shields for your living spaces.

Understanding the intricacies of a well-designed windbreak is paramount for anyone looking to bolster their homestead's defenses against harsh weather conditions. The effectiveness of a windbreak is not merely in its existence but in its strategic planning and execution. Here's a deep dive into crafting a windbreak that stands as a testament to foresight and preparedness:

Height Dynamics: The protective shield offered by a windbreak is fundamentally tied to its height. The rule of thumb in effective windbreak design states that a properly constructed barrier can safeguard an area up to ten times its height. This means a 10-foot tall windbreak could potentially shield a 100-foot stretch from invasive winds. But it's not just about blocking wind; it's about creating a zone where the wind's impact is significantly diminished, providing a safe harbor for your crops, livestock, and home.

Permeability Matters: Counterintuitively, a windbreak with some degree of permeability – around 50-60% – is often more effective than a completely solid one. The reason lies in fluid dynamics; a solid barrier redirects wind upwards, which then descends with turbulent force behind the barrier, potentially causing as much disruption as the initial wind. A semi-permeable windbreak, however, slows the wind without significantly altering its path, spreading out its energy and reducing its velocity in a controlled manner.

Strategic Placement: The positioning of your windbreak should be a calculated decision, taking into account the typical wind directions and the layout of your land. The height of your windbreak directly influences its protective radius; hence, placing it too close to the area you're trying to protect might not provide full coverage, while positioning it too far might render it less effective. Mapping out the prevailing wind patterns and aligning your windbreak accordingly can create a microclimate that shelters sensitive areas of your property.

Layering for Enhanced Protection: The most robust windbreaks employ a layered approach, combining various elements to break down and disperse incoming wind effectively. By integrating both natural elements like trees and shrubs and man-made structures such as fences or walls, you can create a comprehensive barrier that addresses different aspects of wind disruption. The diversity in plant species within natural windbreaks is also key; it ensures year-round protection and minimizes the risks associated with pests and diseases, which can decimate uniform windbreaks.

Regular Upkeep: Like any defensive structure, a windbreak requires ongoing maintenance to retain its effectiveness. This entails regular inspections for potential breaches, pruning to maintain the desired permeability and shape, and repairs to any structural elements. Seasonal checks are advisable to prepare for different weather conditions, ensuring that your windbreak remains robust and intact year-round.

Using physical activity to generate body heat.

Physical activity isn't just good for your health; it's a vital strategy for generating body heat. Simple movements like walking, stretching, or performing jumping jacks can raise your core temperature, making a significant difference in how warm you feel. This is particularly beneficial during power outages or in off-grid scenarios where traditional heating may not be available.

Eating and staying hydrated play essential roles in thermal regulation. Consuming warm, calorie-rich foods and drinks can significantly boost your inner warmth. Think hot soups, stews, or teas which can provide both nutritional energy and an internal source of heat.

However, it's crucial to understand the risks tied to non-traditional heating methods. Be vigilant of the dangers like carbon monoxide poisoning from unventilated heaters or the risk of chemical burns from heat packs. Safety should always remain your top priority, ensuring that the quest for warmth doesn't lead to unintended harm.

This comprehensive guide isn't just about combating cold; it's a strategic approach to maintain comfort and energy efficiency, enhancing your home's warmth without excessive reliance on heating resources. By blending these strategies, you can forge a versatile plan for warmth, guaranteeing comfort and safety even in the face of harsh conditions.

In the realm of sustainable lighting, exploring both cutting-edge and time-tested solutions can revolutionize how you illuminate your space during emergencies or in off-grid living. This segment will delve into practical, efficient lighting methods that ensure you're never left in the dark, regardless of the grid's status. From solar-powered lights to traditional oil lamps, understanding how to utilize various lighting sources can transform your space, providing safety and comfort amidst the darkness.

In emergency preparedness, knowledge is as crucial as the supplies you gather. By equipping yourself with these strategies and understanding their applications, you're not just preparing for potential scenarios; you're ensuring that you and your loved ones can continue living comfortably, regardless of external circumstances, embodying the essence of true self-reliance and adaptability in the face of challenges.

Refocusing on the essence of practical, hands-on advice tailored to self-reliant individuals facing emergencies:

When electricity fails or you're miles from the nearest power line, maintaining visibility is crucial for both safety and normalcy. Efficient lighting means leveraging modern solutions like low-energy LEDs and natural resources to keep your space lit under any circumstances.

Solar power is your silent, relentless ally. Devices that convert sunlight into stored energy can light up your nights without fail. Imagine solar path lights standing guard along your walkway, or solar lanterns hanging in your living room, all charging silently by day to protect and illuminate by night. For indoors, solar-powered LED solutions are invaluable. Charge them by day, and they'll repay you with hours of light after sunset. Consider portable solar panels a worthwhile investment—they're your lifeline to power in remote settings.

LEDs have revolutionized the way we think about light efficiency. They're not just bulbs; they're durable, long-lasting beacons in the night, using a fraction of the energy traditional bulbs consume. Their robust nature means they withstand rough conditions, making them a staple in any emergency kit.

Now, let's get physical with hand-crank and pedal-powered lights. These aren't just gadgets; they're your muscle-powered lifelines in a blackout. Cranking for a few minutes can yield hours of light—a fair trade for invaluable illumination. Pedal generators take this concept further, charging entire batteries or powering multiple lights, a boon in prolonged off-grid scenarios.

Venturing into less conventional territory, consider the untapped potential of bioluminescent lighting. While still largely in the research phase, imagine a future where glowing algae provide ambient light to your home, a blend of science fiction turned science fact.

Wind-up technology mirrors the hand-crank mechanism, providing light with a few turns of a handle. These gadgets often come with extra features like radios or charging ports, making them multifunctional tools in your resilience arsenal.

Gravity-powered lighting, another innovative solution, transforms the pull of gravity into light, without batteries or electricity. Fill a bag with rocks or sand, hoist it up, and as it descends, you're gifted with light—a brilliant off-grid lighting solution.

Reflective and passive lighting techniques should not be underestimated. During daylight hours, maximize natural light with strategically placed mirrors and light-colored walls, reducing the dependency on artificial sources.

When all else fails, candles—especially those made from natural, renewable resources like beeswax or soy—can light up a room. Remember, safety is paramount; never leave candles unattended.

Embracing community projects for sustainable lighting not only enhances your own living conditions but strengthens the entire community's resilience. Initiatives like solar street lamps or shared charging stations foster safety and unity.

Educating your community on these sustainable options empowers and spreads the culture of preparedness and self-reliance. Hosting workshops or sharing knowledge on these lighting solutions can transform individual efforts into collective strength.

Stay abreast of advancements in sustainable lighting. Innovations are constantly emerging, offering new solutions to old problems. Being informed means being prepared.

Incorporate a variety of these lighting strategies into your preparedness plan. Diversity in your lighting sources ensures that, come what may, the light remains on.

Beyond lighting, embracing sustainable practices across all facets of living not only reduces your environmental footprint but fortifies your autonomy. From the food you eat to the way you generate power, every step towards sustainability is a step towards deeper self-reliance.

Understanding your environment, potential threats, and having preemptive measures in place fortifies not just your physical safety, but your peace of mind. Preparedness is more than a set of actions; it's a mindset of readiness, awareness, and continual adaptation, ensuring you and your loved ones remain secure and resilient in any situation.

Fortifying your home is paramount in creating a secure refuge against unforeseen dangers. Start with the basics: strengthen your doors, upgrade your locks, and reinforce windows to thwart unwanted entry. But securing a home doesn't stop at hardening entry points; it extends to creating an integrated defense network. Implement strategic lighting to eliminate dark corners where threats could lurk, erect robust fences for an added layer of protection, and deploy surveillance systems to monitor and deter potential intruders. Leverage cutting-edge technology with alarm systems and motion detectors, ensuring you're alerted at the first sign of intrusion.

Preparation for emergencies is not just a duty; it's a fundamental aspect of living responsibly off-grid. Here's how you ensure readiness:

Stockpile essentials: Maintain a rotating supply of food, water, and medical supplies.

Acquire key skills: Commit to learning and practicing first aid, fire safety, and basic survival tactics.

Draft an evacuation blueprint: Familiarize yourself with all possible exits and plan multiple escape routes from your home.

For self-defense readiness, awareness, and education are your best weapons. Cultivate a keen sense of situational awareness to avoid potential threats before they escalate. Consider acquiring skills in martial arts or self-defense courses as a means to protect yourself and your loved ones. Additionally, familiarize yourself with the legalities and effective usage of personal defense tools such as pepper spray, stun guns, and firearms where applicable and legal.

In today's era, cybersecurity is an extension of personal safety. Protect your online presence with robust passwords, enable two-factor authentication, and ensure your Wi-Fi network is secure. Educate yourself and your family about the dangers of phishing and scams. Regularly updating your devices' software closes security loopholes and safeguards your personal information.

Building a community network is invaluable. Strong relationships with neighbors and local members enhance mutual support and information sharing in times of need. Participating in or establishing community safety initiatives can elevate security for everyone involved.

Remember, personal security is not just physical. Maintaining mental and emotional well-being, particularly in high-stress situations, is crucial. Techniques such as mindfulness, meditation, and staying informed about local risks and emergency procedures bolster both your mental clarity and resilience.

Conduct regular reviews and drills of your emergency plans. This could mean rehearsing evacuation routes, performing lockdown drills, or simulating first-aid scenarios. Adjust and refine your strategies as circumstances change, ensuring your preparedness measures are always up to date.

Adaptability is key in facing new and evolving threats. Commit to continuous learning about advancements in security technology, survival strategies, and health practices. Engage with your community through safety workshops or training sessions to strengthen communal resilience.

Recognize that true security is holistic, encompassing physical safety, mental well-being, and the readiness to act. Integrating these aspects into a comprehensive approach ensures you are equipped to face a wide range of challenges, promoting a safe, prepared, and resilient lifestyle.

By adopting a thorough approach to personal security, you not only safeguard yourself and your loved ones but also contribute to the strength and preparedness of your wider community. This commitment transcends immediate concerns, laying the groundwork for sustained security and peace of mind.

In navigating the complexities of self-defense, balance is essential. Equip yourself with knowledge and tools, but always weigh the moral and legal implications of your actions. Self-defense is a right, yet it demands responsibility — understanding the fine line between necessary force and excess is critical.

Case Study: The Thompson Family Secure Homestead.

Consider the Thompson family, living on a secluded homestead vulnerable to natural disasters and isolated incidents. They applied a multifaceted approach to security:

Fortification: They upgraded all doors to steel-core versions and installed double-pane, shatter-resistant windows.

Layered Defense: Motion-sensitive floodlights were mounted around the perimeter, a sturdy fence was erected, and multiple, discreet surveillance cameras were installed.

Skill Development: Each family member attended first aid and fire safety workshops, ensuring collective preparedness.

Their efforts were put to the test during a significant wildfire. With roads blocked and communications down, their preparations paid off. Their home resisted the fire's advance, and their supplies and skills provided sustenance and safety until they could safely evacuate.

This real-life example underscores the efficacy of a holistic security strategy, combining physical reinforcements, skill acquisition, and community ties. By mirroring the Thompsons' approach, you can significantly enhance your and your family's safety, ensuring preparedness for various emergencies.

Before you even think about adding any kind of weapon to your home defense plan, it's critical to stop and consider the ethical side of things. Reflect deeply on what it means to bring such tools into your life. Think about the chance of accidental injuries, the heavy responsibility that comes with owning and potentially using a weapon, and how it could affect you mentally if you ever had to use force to defend yourself or your family.

Dive into the legal side of things next. Each place has its own rules about defending yourself and what kind of tools you can legally use to do so. Get to know these laws inside out. Understanding what's considered reasonable force in your area is crucial to ensure you act within the law should you ever need to defend yourself.

When it comes to choosing tools for self-defense, don't just pick something off the shelf without thinking. Consider how easy it is to use, how dependable it is, and what the law says about it. You've got a range of options, from non-lethal items like pepper spray or stun guns to more serious tools like firearms. Each requires a different level of training and understanding of legal rules.

Effective self-defense is all about skill, not just the tools you have. You need to know how to use these tools properly, especially when under pressure. This means regular, real-world training is a must. This ensures you're ready and able to use whatever tools you choose effectively and responsibly if the situation arises.

Staying out of trouble in the first place is a big part of self-defense. This means being switched on about what's happening around you, spotting potential threats early, and knowing when it's time to get out of a situation before it gets worse.

But remember, the best way to defend yourself isn't always with physical force. Learning how to calm things down and talk through problems can often stop conflicts before they start. Effective communication and knowing how to de-escalate a tense situation can be your best defense tools.

Make your home a fortress, but in a smart way. Integrate your self-defense plans into your overall home security strategy. Secure your doors and windows, install surveillance cameras, and have a solid emergency plan in place. This all-round approach boosts your safety significantly.

Don't forget the mental side of things. Facing a self-defense situation can leave marks that aren't just physical. Be aware of the mental and emotional impact such events can have and have support systems and strategies in place to help you cope afterward.

In adopting these strategies, your home transforms from a simple dwelling into a fortified bastion.

This transformation is not merely for survival; it's a proactive step towards a life of safety, self-reliance, and readiness. Your commitment to home defense is ongoing, adapting as new threats emerge. Consistently review and enhance your security measures to ensure your haven stands resilient.

The real strength of your fortress lies in knowledge, preparation, and community. Arm yourself with information, equip your home with necessary defenses, and stock your pantry for the long haul. Build relationships with neighbors and local law enforcement to strengthen your security network.

Your home is more than a shelter; it's a reflection of your commitment to protect and sustain your way of life. Let this guide be a starting point for a lifestyle steeped in preparedness. Stand confident, knowing your fortress is secure, prepared for whatever lies ahead. Welcome to true security. Welcome to your fortress.

Book 5: Provisions for the Long Haul - Mastering Food Security

When diving into long-term food storage, envision preparing your pantry like it's a lifeline for future you. It's about ensuring that, come what may, you have a supply of food that's safe, nutritious, and tasty years down the line. Think of it as insurance in your basement – food that's ready whenever you need it, ensuring peace of mind and sustenance through whatever comes your way.

Long-term food storage isn't just about preparing for the worst; it's about securing your well-being in uncertain times. Natural disasters, economic fluctuations, or situations where access to stores is restricted can all leave you vulnerable if you haven't stocked up. Your food stash becomes your lifeline, ensuring you can weather any storm without going hungry.

Not all foods are created equal when it comes to long-term storage. To build a robust food supply, focus on items low in moisture and high in stability. Grains like wheat, rice, and oats, along with hardy legumes such as beans and lentils, are excellent choices. Essentials like powdered milk, sugar, and salt are also key components, while honey, dried fruits, and veggies shine as longevity stars. Stored properly, these foods can remain edible for years, providing sustenance when you need it most.

Air, moisture, light, and unwanted critters are the enemies of long-term food storage. These culprits can quickly turn your food supply from a feast to fodder if left unchecked. Combat them by ensuring your storage containers are airtight and moisture-proof. Keep your supplies in a cool, dark location to prevent spoilage and degradation. Opt for pest-proof containers to protect against insects and rodents. Regular checks and rotation of your food stocks are essential to maintaining their quality and edibility over time.

Temperature plays a crucial role in preserving the longevity of your stored food. Cooler, stable conditions are ideal, as they slow down the natural degradation processes that can occur over time. Consider your basement or a root cellar as the VIP lounge for your long-term storage. These areas provide the consistent, low temperatures needed to keep your food fresh and viable for extended periods. By storing your food in the right conditions, you ensure it remains ready and edible, just like the day you stored it.

Long-term food storage is a vital component of emergency preparedness for practical, self-reliant individuals. By stocking up on the right foods and storing them properly, you can safeguard your well-being and ensure you're prepared for whatever challenges come your way. Take the time to build a robust food supply, combatting the enemies of storage and leveraging the power of temperature to preserve your provisions. With a well-stocked pantry and a solid plan in place, you can face the future with confidence, knowing you have the sustenance you need to thrive in any situation.

Moisture's the ninja of food spoilers, creeping into your stash and causing chaos. Tools like silica gel packets and oxygen absorbers are your allies in this fight, keeping your goods dry and safe. Regular vigilance ensures your food remains untouched by moisture's stealthy advances.

And pests? They're the opportunists, always on the prowl for a chance to infiltrate your supplies. The best defense against these invaders is to store your food in strong, impenetrable containers and to keep your storage areas clean and orderly. Conducting regular checks allows you to catch any issues early, preventing a small problem from becoming a full-blown infestation.

Weevils, for instance, are tiny beetles that can lay eggs in grains like rice and flour. These eggs hatch into larvae that feed on the grains, contaminating them with their waste and rendering them unfit for consumption. Ants are another common nuisance, especially in warmer climates. They can infiltrate your storage area in search of sugary or protein-rich foods, quickly establishing colonies and causing widespread damage.

To deter these pests, consider adding natural repellents such as bay leaves, cloves, or diatomaceous earth to your storage containers. Bay leaves contain compounds that repel insects, while cloves emit a strong aroma that can deter pests from approaching your food. Diatomaceous earth is a fine powder made from fossilized algae that acts as a natural insecticide, dehydrating and killing insects upon contact. These substances can help repel insects and prevent infestations from taking hold.

Rodents pose another significant threat to your food supplies, with their sharp teeth and relentless determination to find a meal. Rats and mice can chew through plastic containers and cardboard boxes with ease, gaining access to your stored food and leaving behind droppings and urine that can contaminate it. To keep rodents at bay, ensure that your storage areas are well-sealed and free of any potential entry points. Use sturdy containers made of metal or heavy-duty plastic, and elevate them off the ground to prevent access from rodents and other critters.

Regular checks are essential for detecting and addressing pest issues before they spiral out of control. Keep a vigilant eye on your stored food, inspecting containers for signs of damage or infestation. Look for telltale signs such as chew marks, droppings, or strange odors that may indicate the presence of pests.

Take immediate action to contain the problem by removing and disposing of any contaminated items. Clean and sanitize your storage containers thoroughly to eliminate any lingering odors or attractants that may attract pests. Consider investing in traps or baits to control the pest population and prevent further damage to your supplies.

By staying vigilant and proactive in your pest management efforts, you can protect your long-term food storage from the threat of infestation. With proper storage techniques and regular maintenance, you can safeguard your supplies and ensure they remain safe, secure, and ready to sustain you in times of need.

Incorporating these elements into your long-term food storage strategy turns your pantry into more than just a collection of food; it becomes a well-maintained reserve, ready to support you and your family through times of need. This isn't about hoarding; it's about smart, strategic planning and execution to ensure you're never caught unprepared, regardless of what life throws your way.

In the world of preparedness and self-sufficiency, routine inspections of your food storage are as critical as the initial act of storing. This isn't a mere set-and-forget scenario; it's akin to routine reconnaissance of your supply lines. Periodically survey your cache for any signs of wear, spoilage, or pest intrusion. Implement the "first in, first out" methodology, cycling through older stock to ensure freshness and viability. This strategy not only maximizes your resources but also fortifies your readiness, ensuring that every can, every bag, every grain remains a viable asset.

In times of crisis or uncertainty, having a reliable food supply is paramount for survival. Canning is a time-tested method of preserving food that allows you to stockpile nutritious staples for extended periods. In this guide, we'll delve into the art of canning, providing you with the knowledge and skills to safely and effectively preserve food for long-term storage.

Understanding Canning.

Canning is a process of preserving food in airtight containers, typically glass jars, to prevent spoilage and maintain freshness. By sealing food in jars and subjecting them to heat, bacteria and other microorganisms are destroyed, allowing the food to be stored for an extended period without refrigeration.

Water bath canning is suitable for high-acid foods such as fruits, pickles, and jams. It involves submerging sealed jars in boiling water for a specified period to kill bacteria and create a vacuum seal.

To water bath can, start by preparing your ingredients and sterilizing your jars, lids, and utensils. Fill the jars with the prepared food, leaving the appropriate headspace, and then seal them with lids and rings. Submerge the jars in boiling water using a canning rack, ensuring they are fully covered with water, and process them for the recommended time according to the recipe.

Case Study: Emily, a homesteader living in a rural area, decided to water bath can a surplus of peaches from her orchard. Following proper canning procedures and processing the jars for the specified time, she successfully preserved jars of delicious peach preserves that would last for years.

Pressure canning is necessary for low-acid foods such as vegetables, meats, and soups, as well as high-acid foods at altitudes above 1,000 feet. It involves using a pressure canner to heat jars to a higher temperature than boiling water, which is necessary to kill harmful bacteria like Clostridium botulinum.

To pressure can, begin by preparing your ingredients and sterilizing your jars, lids, and pressure canner. Fill the jars with the prepared food, leaving the appropriate headspace, and then seal them with lids and rings. Place the jars in the pressure canner with water, following the manufacturer's instructions for venting and pressurizing. Process the jars for the recommended time and pressure based on the type of food and your altitude.

Case Study: Tom, a prepper living in a mountainous region, decided to pressure can a variety of vegetables from his garden. By carefully following the guidelines for pressure canning and adjusting the processing time and pressure for his altitude, he successfully preserved jars of nutrient-rich vegetables to sustain his family through the winter months.

Understanding Dehydrating:

Dehydrating is a practical and effective method for extending the shelf life of food while retaining its nutritional value. In this guide, we'll explore the ins and outs of dehydrating, equipping you with the knowledge and skills to preserve a variety of foods for long-term storage.

Dehydrating, also known as drying, involves removing moisture from food to inhibit the growth of bacteria, mold, and yeast. By reducing the water content, food becomes lightweight, compact, and shelf-stable, making it ideal for long-term storage without the need for refrigeration.

When it comes to dehydrating food, investing in a quality dehydrator is essential. Consider factors such as size, temperature control, and airflow when selecting a dehydrator that suits your needs. Look for models with adjustable temperature settings and ample drying space to accommodate various types of food.

Before dehydrating food, it's crucial to properly prepare and pre-treat the ingredients to ensure optimal results. Wash and slice fruits and vegetables into uniform pieces to promote even drying. Blanching or steam blanching certain vegetables before dehydrating can help preserve color and texture while reducing enzyme activity.

Fruits are excellent candidates for dehydration, as they retain their natural sweetness and flavor when dried. The most popular fruits for dehydration are apples, bananas, berries, and mangoes. To dehydrate fruit, arrange slices on dehydrator trays in a single layer and set the temperature according to the specific fruit's requirements. Monitor the drying process regularly and rotate trays as needed for even drying.

Dehydrated vegetables are versatile ingredients that can be rehydrated for use in soups, stews, and casseroles. Common vegetables for dehydration include carrots, peppers, onions, and tomatoes. Prepare vegetables by blanching or steam blanching before arranging them on dehydrator trays. Dry vegetables at a low temperature until they are crisp and brittle.

Herbs are simple to dehydrate and add flavor and aroma to your culinary creations. Harvest fresh herbs from the garden and wash them thoroughly before removing any excess moisture. Arrange herbs in a single layer on dehydrator trays and dry at a low temperature until they are thoroughly dried and crumbly.

Case Study: Meet Sarah, a self-reliant homesteader with a passion for preserving food. Concerned about food security, Sarah decided to invest in a quality dehydrator to bolster her emergency food supply. Using surplus produce from her garden, Sarah dehydrated a variety of fruits, vegetables, and herbs. By following proper dehydration techniques and storing the dried food in airtight containers, Sarah ensured she had a nutritious and shelf-stable food reserve for any emergency.

Dehydrating is a valuable skill for self-reliant individuals seeking to build a long-term food supply. By mastering the art of dehydration, you can preserve a wide range of foods for emergency use, ensuring you have access to nutritious ingredients when needed most. Start dehydrating today and take control of your food security journey.

Freeze drying is a highly effective method for preserving food, offering unparalleled shelf life and retention of flavor and nutrients. In this guide, we'll delve into the world of freeze drying, providing practical advice and actionable steps to help you preserve a variety of foods for long-term storage.

Understanding Freeze Drying:

Freeze drying, also known as lyophilization, involves freezing food at ultra-low temperatures and then removing moisture through sublimation. This process results in lightweight, shelf-stable food with minimal loss of flavor, texture, and nutritional content. Freeze-dried food retains its original shape, color, and taste, making it an ideal option for emergency food storage.

While freeze drying equipment can be costly, it is a worthwhile investment for those serious about long-term food preservation. When choosing a freeze dryer, consider factors such as capacity, performance, and ease of use. Look for models with adjustable settings and ample drying space to accommodate a variety of foods.

Before freeze drying food, it's essential to properly prepare and pre-treat the ingredients to ensure optimal results. Wash and slice fruits and vegetables into uniform pieces, and blanch or steam blanch certain vegetables to preserve color and texture. For meats, trim excess fat and slice thinly for faster drying.

Freeze-dried fruits retain their natural sweetness and vibrant color, making them a delightful addition to your emergency food supply. Prepare fruits by slicing them into uniform pieces and arranging them on freeze dryer trays in a single layer. Freeze dry fruits at low temperatures until they are crisp and dry.

Freeze-dried vegetables are lightweight and versatile, perfect for adding nutrition to soups, stews, and casseroles. Prepare vegetables by blanching or steam blanching before arranging them on freeze dryer trays. Freeze dry vegetables until they are brittle and moisture-free.

Freeze-dried meats provide a convenient source of protein for emergency meals. Prepare meats by slicing thinly and removing excess fat. Arrange meat slices on freeze dryer trays and freeze dry at low temperatures until they are fully dehydrated. Store freeze-dried meats in airtight containers for long-term storage.

Freeze drying is a highly effective method for preserving food, offering unparalleled shelf life and retention of flavor and nutrients. By mastering the art of freeze drying, you can create a diverse and nutritious food supply that will sustain you and your loved ones through any crisis. Start freeze drying today and take control of your food security journey.

The natural world offers an open-air market of sustenance for those versed in the arts of hunting, fishing, and foraging. Viewing the great outdoors through the lens of utility and sustenance brings us closer to our roots and prepares us for a future where the luxury of choice may be a memory. This knowledge is power; it's freedom in its purest form, enabling you to feed, heal, and sustain yourself directly from nature's bounty.

Hunting for food.

Every one of us has a primal instinct—an innate connection to the earth waiting to be awakened. Whether you find yourself amidst dense forests, sprawling deserts, or by a murmuring stream, opportunities to source nourishing food abound. This section is dedicated to rekindling that connection and enabling you to tap into nature's bounty just like our ancestors did.

Before we delve into the specifics, let's discuss responsibility. Ethical hunting and fishing entail taking only what you need, abiding by wildlife laws, and recognizing the significance of conservation efforts. It's about maintaining balance—ensuring that our survival skills do not jeopardize the delicate ecosystems we rely on.

Choose a firearm that is appropriate for the game you intend to hunt. Rifles are commonly used for larger game such as deer, while shotguns are effective for birds and small game. Ensure your firearm is properly maintained and sighted in before heading into the field.

If you prefer archery hunting, select a bow that suits your shooting style and draw weight. Practice regularly to improve accuracy and proficiency with your bow. Consider using broadheads specifically designed for hunting to ensure clean and ethical kills.

Choose the right dress for your environment. Opt for camouflage clothing that blends into your surroundings and provides protection from the elements. Don't forget essential gear such as binoculars, a knife, and a first aid kit.

Hunting isn't merely about the pursuit; it's a profound understanding of animal behavior, adept tracking skills, and utilizing tools with precision. It demands patience, reverence for nature, and relentless practice.

Scouting is a fundamental aspect of successful hunting, providing valuable insights into the terrain, animal behavior, and potential hunting opportunities. Here's how you can effectively scout your hunting area:

Start by researching your hunting area using maps, satellite imagery, and local knowledge. Identify key features such as water sources, food plots, and natural funnels where game is likely to travel.

Utilize online resources and hunting forums to gather information on recent sightings, animal movements, and hunting pressure in the area.

Once on-site, explore the terrain on foot to familiarize yourself with the landscape and locate potential hunting spots. Look for game trails, bedding areas, and feeding areas where animals congregate.

Use binoculars or a spotting scope to scan the area for signs of wildlife activity, such as tracks, droppings, and rubs on trees.

Deploy trail cameras in strategic locations to monitor animal movements and behavior. Set up cameras along game trails, near water sources, and at pinch points to capture images of passing wildlife. Check the trail camera footage regularly to gather valuable data on the size, sex, and behavior of the game species in your hunting area.

Stealth is essential for getting close to game animals without alerting them to your presence. Here are some stealth techniques to improve your hunting success:

Slow down your movements and take deliberate steps to minimize noise. Avoid snapping twigs, rustling leaves, or making sudden movements that could startle nearby game.

Use stealthy footwear with soft soles or specialized hunting boots designed to dampen noise and provide traction on various terrains.

Take advantage of natural cover such as trees, bushes, and rock formations to break up your silhouette and blend into your surroundings. Use shadows and terrain features to remain concealed from the keen eyesight of game animals.

Consider using a ground blind or tree stand to elevate your position and increase your field of view while remaining hidden from sight.

Minimize human scent by showering with scent-free soap and using scent-eliminating sprays or clothing. Store hunting gear in scent-proof containers and avoid contaminating your hunting area with foreign odors.

Pay attention to wind direction and plan your approach to keep your scent from drifting towards game animals. Position yourself downwind of your target to reduce the risk of detection.

Tracking is a valuable skill that allows hunters to follow game trails and locate animals. Here's how you can improve your tracking abilities:

Learn to recognize and interpret animal sign such as tracks, scat, hair, and feeding patterns. Look for fresh tracks in soft soil or snow and examine the size, shape, and direction of travel to determine the species and age of the track.

Use tracking guides and reference books to familiarize yourself with the tracks and signs of common game species in your hunting area.

Practice aging tracks by observing changes in track shape, clarity, and moisture content over time. Fresh tracks will have crisp edges and distinct features, while older tracks may be weathered, filled with debris, or partially obscured.

Use tracking sticks or your hand to measure the size of tracks and estimate the age of the animal based on the depth of the impression and environmental conditions.

Follow game trails methodically, scanning the ground for signs of passage and paying attention to subtle clues such as broken branches, disturbed vegetation, and droppings. Use a slow and deliberate pace to avoid overlooking important indicators.

Keep your eyes focused ahead while tracking, but periodically stop to listen for sounds of movement or calls from nearby animals. Use binoculars or a monocular to scan the surrounding area for any signs of wildlife activity.

Case Study: John, an experienced hunter, decided to scout a new hunting area before the start of deer season. Using topographic maps and satellite imagery, John identified several promising locations with thick cover, natural funnels, and nearby water sources. He deployed trail cameras along game trails and set up a ground blind overlooking a well-used bedding area. Employing stealth techniques, John moved quietly and concealed himself effectively, allowing him to observe deer behavior without detection. Using his tracking skills, John followed fresh tracks to locate a buck feeding in a nearby clearing, ultimately leading to a successful harvest.

Fishing

Fishing is a valuable skill that provides a renewable and sustainable food source, especially in areas with accessible water bodies.

Fishing for survival involves using various techniques to catch fish for sustenance when other food sources may be limited or unavailable. Whether you're stranded in the wilderness or facing a prolonged emergency situation, knowing how to fish can provide you with a reliable source of protein and nutrients.

Selecting the appropriate fishing equipment is crucial for successful fishing in survival situations.

Choose a fishing rod and reel combination that suits your intended fishing environment, whether it's freshwater streams, lakes, or coastal waters. Ensure your equipment is durable, lightweight, and portable, making it suitable for inclusion in a survival kit or bug-out bag.

Consider factors such as rod length, action, and reel type based on your fishing preferences and the target fish species. Pack a variety of fishing line, hooks, sinkers, and lures to adapt to different fishing conditions and bait preferences. Additionally, include basic tackle repair tools such as pliers, scissors, and extra line to address any equipment malfunctions in the field.

When it comes to fishing for survival, adaptability and resourcefulness are key. Explore a range of fishing techniques and strategies to increase your chances of success in various environments and conditions.

Shore fishing is a versatile and accessible method for catching fish from the shoreline of rivers, lakes, or coastal areas. Look for areas with natural cover, such as rocks, fallen trees, or vegetation, where fish are likely to congregate. Cast your line near these features and vary your bait presentation to entice fish. Experiment with different bait options, including natural bait such as worms, insects, or small fish, as well as artificial lures designed to mimic prey movements. Use a bobber or float to suspend bait at different depths and cover a larger area of the water column. Adjust your fishing tactics based on water conditions, weather patterns, and the behavior of feeding fish.

Fly fishing is an elegant and effective method for catching freshwater fish species such as trout, salmon, and bass. Practice casting and presenting flies accurately to mimic insect activity and attract fish to strike. Match your fly selection to the local insect hatch and fish preferences, focusing on natural patterns and sizes that closely resemble the available prey.

In a case study, let's consider John, an experienced outdoorsman and survivalist who found himself stranded in a remote wilderness area due to unforeseen circumstances. With limited food supplies and no immediate means of rescue, John relied on his fishing skills to sustain himself until help arrived. Using a simple fishing rod, tackle box, and improvised bait, John caught several fish from a nearby stream, providing him with much-needed sustenance and energy during his ordeal.

Fishing for survival is a valuable skill that can provide sustenance and nourishment in emergency situations. By mastering fishing techniques and selecting the right equipment, you can increase your chances of successfully catching fish for food. Start honing your fishing skills today and take control of your food security and survival preparedness journey.

Understanding Foraging.

In a world where self-reliance is paramount, mastering the art of foraging can be a game-changer. Whether you're faced with a survival situation or simply looking to supplement your pantry with wild edibles, knowing how to identify and harvest edible plants can provide a valuable source of nutrition. In this guide, we'll delve into the essentials of foraging, offering actionable advice and real-world applications for practical, hands-on individuals focused on self-reliance in emergencies.

Foraging is the practice of gathering wild food resources, including edible plants, berries, nuts, and mushrooms, from their natural environment. Before venturing into the wild, it's crucial to familiarize yourself with local flora and fauna, as well as foraging ethics and safety considerations. Armed with knowledge and caution, you can unlock the bounty of nature and sustain yourself with nutritious, wild-harvested foods.

The key to successful foraging lies in the ability to identify edible plants accurately. Start by learning to recognize common edible species in your region, focusing on characteristics such as leaf shape, color, texture, and growth habitat. Field guides, online resources, and local foraging workshops are valuable tools for expanding your botanical knowledge and honing your plant identification skills.

Once you've identified edible plants, it's time to harvest and prepare them for consumption. Use a sharp knife or scissors to carefully collect plant parts such as leaves, stems, flowers, and roots, taking care to leave behind enough foliage for the plant to continue growing. Avoid harvesting from polluted areas or plants that may be contaminated with pesticides or other chemicals.

After harvesting, thoroughly wash and inspect the wild edibles to remove any dirt, insects, or other debris. Depending on the plant species, you may need to blanch, sauté, steam, or consume them raw. Experiment with different cooking methods and flavor combinations to discover your favorite ways of incorporating wild edibles into meals and snacks.

Foraging comes with inherent risks, including the potential for misidentification and accidental ingestion of toxic plants. Always cross-reference multiple sources and consult with experienced foragers or botanists to verify plant identities before consuming them.

Avoid harvesting from areas that may be contaminated with pollutants or industrial runoff, such as roadside ditches or heavily sprayed fields. Additionally, be mindful of seasonal variations in plant appearance and availability, as well as any regulations or restrictions on foraging in public lands or protected areas.

Tina and Mark, avid outdoor enthusiasts, embarked on a weekend hiking adventure in the remote wilderness of the Pacific Northwest. Excited to explore the beauty of the region, they set off with a well-packed backpack and a sense of adventure.

As they ventured deeper into the wilderness, they encountered challenging terrain and unpredictable weather conditions. Despite their best efforts to stay on course, they soon realized they were lost and disoriented, with no clear path back to civilization.

With nightfall approaching and their supplies dwindling, Tina and Mark knew they needed to take decisive action to ensure their survival. Drawing on their knowledge of wilderness survival techniques and years of outdoor experience, they assessed their situation and formulated a plan. Relying on their foraging skills, they scoured the surrounding landscape for edible plants, berries, and wild mushrooms. Drawing from their knowledge of local flora and fauna, they identified safe and nutritious options to supplement their dwindling food supply.

Each day became a new opportunity for discovery as Tina and Mark explored their surroundings, learning to appreciate the abundance of nature and the interconnectedness of all living things. They marveled at the resilience of the wilderness and the inherent beauty of the natural world, finding solace and inspiration in their surroundings.

Despite the challenges they faced, they remained resilient and optimistic, drawing strength from each other and their shared love of the outdoors. They supported one another through moments of doubt and uncertainty, finding comfort in their mutual determination to persevere.

As days turned into weeks, Tina and Mark's resourcefulness and ingenuity were put to the test. They constructed makeshift shelters from fallen branches and debris, using their survival skills to adapt to the ever-changing conditions of the wilderness.

Their perseverance paid off when, after weeks of wandering, Tina and Mark stumbled upon a remote hunting cabin nestled deep in the forest. Overjoyed at their stroke of luck, they sought refuge in the cabin, finding shelter, warmth, and safety in the midst of the wilderness.

Tina and Mark's uplifting survival story serves as a testament to the resilience of the human spirit and the power of perseverance in the face of adversity. Through their resourcefulness, ingenuity, and unwavering determination, they not only survived but thrived in the wilderness, emerging stronger, wiser, and more connected to the natural world than ever before.

Foraging is a valuable skill that offers both practical and philosophical benefits for self-reliant individuals seeking to connect with their environment and sustain themselves in times of need. By learning to identify, harvest, and prepare wild edibles responsibly, you can enhance your resilience and resourcefulness while enjoying the abundance of nature's pantry. Start exploring the world of foraging today and unlock the hidden treasures of the wilderness.

Sharing knowledge and experiences can enrich your journey immeasurably. Engage with local foraging groups, hunting clubs, and fishing communities to glean valuable insights and garner support. This section underscores the significance of connecting with these communities and underscores the importance of passing on your acquired wisdom.

In the realm of emergency preparedness, knowledge truly is power. As we've explored in this comprehensive guide, mastering the art of long-term food storage, revolutionary food preservation techniques, and harnessing the abundance of nature through hunting, fishing, and foraging are essential skills for self-reliance in uncertain times.

From stocking our pantries with shelf-stable staples to unlocking the secrets of preserving any type of food, we've equipped ourselves with the tools and know-how to thrive in the face of adversity. By embracing the age-old practices of our ancestors and blending them with modern innovations, we've unlocked the potential to sustain ourselves and our loved ones for years to come.

But our journey doesn't end here. In fact, it's only just begun. As we continue to hone our skills and expand our knowledge, we're reminded of the importance of preparedness as a way of life. It's about more than just stockpiling supplies; it's about cultivating a mindset of resilience, resourcefulness, and adaptability in the face of whatever challenges may arise.

As we navigate an increasingly uncertain world, we take comfort in knowing that we have the tools, techniques, and wisdom to weather any storm. From the pantry shelves lined with carefully preserved provisions to the wilderness teeming with nature's bounty, we stand ready to embrace the challenges of tomorrow with confidence and determination.

So, let us carry forward the lessons learned and the skills acquired on this journey of preparedness. Let us continue to educate ourselves, to practice our craft, and to share our knowledge with others. For in doing so, we not only empower ourselves but also uplift and inspire those around us to embrace the ethos of self-reliance and resilience.

May we remain vigilant, may we stay prepared, and may we always be ready to face whatever challenges come our way. With dedication, determination, and a steadfast commitment to self-reliance, we can navigate the uncertainties of tomorrow with strength, courage, and unwavering resolve.

Book 6: Communicating When Everything Else Fails

When it comes to preparing for emergencies, one of the most critical aspects is ensuring effective communication. Let's take a moment to consider your unique communication needs in times of crisis. Picture yourself in an emergency situation - perhaps a natural disaster, power outage, or civil unrest. Who are you responsible for? Your family? Friends? Neighbors? Assessing the size and composition of your household or group is the first step in understanding your communication needs.

Now, let's think about the nature of potential emergencies you might face. Are you in an urban, suburban, or rural area? Each setting comes with its own set of challenges and communication limitations. For instance, urban areas may have more reliable cell phone coverage but also face the risk of network congestion during emergencies. On the other hand, rural or remote areas might lack cell phone reception altogether, requiring alternative communication methods.

It's essential to recognize the limitations of traditional communication methods in emergency situations. While modern technology has vastly improved our ability to stay connected, it's not infallible. Damage to infrastructure, such as communication towers and power lines, can disrupt traditional networks during disasters. Overload on cell phone networks can also occur when everyone is trying to make calls at once. Additionally, power outages can render devices like landline phones and internet routers useless.

Given these limitations, it's crucial to identify scenarios where off-grid communication becomes essential. Imagine being in a situation where traditional methods are unavailable or ineffective. This could be during a natural disaster when infrastructure is damaged, or in a remote wilderness area with no cell phone coverage. Off-grid communication also becomes vital in emergencies requiring stealth or security, such as home invasions or hostage situations. And of course, in extreme survival scenarios where you're stranded or isolated, the ability to signal for help or communicate with rescuers can mean the difference between life and death.

So, now that you've assessed your communication needs and identified scenarios where off-grid communication is essential, let's explore the options available to you. Whether it's handheld radios, satellite phones, or creative signaling methods, there are solutions to ensure you can stay connected and informed, no matter what challenges come your way.

Off-grid Communication.

When it comes to off-grid communication, there's a wide array of methods to consider, each with its own set of pros and cons. Let's dive into the options available and explore how to choose the best one for your needs.

First up, let's talk about radio communication. Radios have long been a staple in emergency communication, offering reliable, long-range communication without the need for infrastructure. Whether it's a handheld radio, a mobile unit for your vehicle, or a base station for your home, radios provide a versatile and dependable way to stay connected. They're particularly well-suited for communicating in remote or rugged terrain where cell phone coverage is sparse or nonexistent. Plus, radios operate on different frequency bands, allowing you to choose the one that best suits your needs and regulatory requirements.

Next, let's consider satellite communication. Satellite phones, messengers, and trackers offer global coverage, making them ideal for staying connected in remote areas far from civilization. They work by connecting to satellites orbiting the Earth, providing a lifeline when traditional communication methods are unavailable. Satellite communication devices are compact, portable, and relatively easy to use, making them a valuable addition to any emergency kit. However, satellite services typically come with subscription fees and may require a clear line of sight to the sky, limiting their effectiveness in densely wooded or urban environments.

Lastly, let's explore mesh networks. Mesh networks are decentralized communication systems that rely on interconnected nodes to relay messages. Unlike traditional networks that depend on centralized infrastructure, mesh networks can operate independently, making them resilient in the face of disasters or disruptions. Mesh devices, such as portable Wi-Fi routers or mesh radio systems, create a network of interconnected nodes that can communicate with each other even if some nodes are offline or out of range. This flexibility makes mesh networks particularly well-suited for community-based communication and emergency response efforts.

Now, let's talk about the pros and cons of each communication option. Radios are reliable and easy to use, but their range is limited by factors like terrain and antenna height. Satellite communication offers global coverage, but it can be expensive and may require a clear line of sight to the sky. Mesh networks are decentralized and resilient, but they require multiple nodes to function effectively and may have limited range compared to other options.

When evaluating the suitability of different communication methods for your situation, consider factors such as your location, the nature of potential emergencies, and your budget. First of all, are you in a rural or an urban area? Do you need global coverage or just local communication? What are your regulatory requirements and licensing restrictions? By carefully considering these factors, you can choose the off-grid communication method that best meets your needs and ensures you stay connected when it matters most.

Ready to dive into the world of radio communication? It's like having your own personal lifeline when the grid goes down and traditional communication methods fail. Let's break down the essentials in a way that's straightforward and practical, just like you prefer.

So, what's the deal with radio communication? Well, imagine it as your own secret language with the power to reach out and connect with others when you need it most. But before we start talking in radio jargon, let's cover the basics.

First off, let's talk about frequency bands. Think of these like radio channels on your car stereo - each band has its own set of frequencies for communication. We're talking Very High Frequency (VHF), Ultra High Frequency (UHF), and High Frequency (HF). Each band has its strengths and weaknesses, so it's crucial to choose the right one based on your location, terrain, and the kind of communication you need.

Now, let's get into modulation types. This is where it gets a bit technical, but stick with me! Modulation is how we add information to our radio waves. It's like Morse code for the modern age. We've got Amplitude Modulation (AM), Frequency Modulation (FM), and Single Sideband (SSB). Each one has its own quirks and best uses, so it's worth getting to know them like old friends.

And let's not forget about transmission modes. This is all about how we talk back and forth over the airwaves. You've got simplex mode for one-way communication, half-duplex for two-way chatting (one person talks while the other listens), and full-duplex for those full-blown conversations where both parties can talk and listen simultaneously.

Now, onto the fun part - choosing your radio gear! Picture yourself with a trusty handheld radio, AKA the walkie-talkie of your dreams. It's perfect for those short-range chats with your buddy across the field or coordinating with your team during a crisis. Then there's the mobile radio, the powerhouse of your vehicle, giving you more range and oomph for communication on the move. And let's not forget about the base station unit - your home's communication hub, ready to keep you connected when the world outside goes dark.

Here's the nitty-gritty of establishing off-grid communication protocols - a crucial step for anyone serious about self-reliance in emergencies. So, picture this: you're in the midst of a crisis, and traditional communication methods are out the window. What do you do? That's where a solid communication plan comes into play.

First things first, you need to develop a communication plan tailored to your household or group. Sit down with your family, friends, or fellow preppers and discuss your communication needs and priorities. Identify key contacts both within and outside your group, and establish backup communication channels in case primary methods fail. Consider factors like range, reliability, and ease of use when choosing communication devices and methods.

Once you've got your plan in place, it's time to assign roles and responsibilities. Who's going to be the primary communicator? Who's responsible for monitoring incoming messages and relaying information to the rest of the group? Assigning clear roles ensures that everyone knows what they need to do in an emergency and prevents confusion or duplication of efforts.

Now, let's talk about protocols for message encryption, authentication, and verification. In a world where privacy and security are paramount, it's essential to protect your communications from prying eyes and malicious actors. Consider using encryption software or techniques to scramble your messages and prevent interception or tampering. Authenticate the identity of message senders to ensure that you're communicating with trusted sources, and verify the integrity of received messages to guard against misinformation or manipulation.

But wait, there's more! Don't forget to establish protocols for prioritizing messages based on urgency and importance. In an emergency situation, not all messages are created equal. Develop a system for categorizing and triaging incoming messages to ensure that critical information gets through quickly and efficiently.

Now, let's bring it all together with a case study. Meet the Johnson family - a group of self-reliant folks living off the grid in rural Montana. With wildfires raging nearby, they know that communication is key to their safety and survival. They've developed a communication plan that includes a mix of handheld radios, satellite phones, and backup signaling devices. Dad, a former military communications specialist, takes on the role of primary communicator, while Mom is responsible for monitoring incoming messages and relaying information to the kids. They've implemented encryption software on their devices and regularly practice sending and receiving encrypted messages to ensure everyone is familiar with the process. When a wildfire threatens their homestead, the Johnsons spring into action, using their communication protocols to coordinate evacuation efforts and stay informed about changing conditions.

We need to spend some time focusing on the unsung heroes of off-grid communication: handheld radios. When phones fail and traditional methods falter, these gadgets become our lifelines. But with a myriad of options available, how do we select the perfect one? Let's break it down.

Exploring the diverse handheld radio options, we find FRS radios, ideal for short-range communication within a few miles. These are perfect for keeping in touch with loved ones around the home or coordinating activities during outdoor pursuits. Then, there are GMRS radios, offering increased power and range compared to FRS radios. With GMRS, you can communicate over longer distances, making them indispensable for adventures or connecting with family dispersed across a wider area.

Lastly, we can't overlook ham radios - the Swiss Army knives of off-grid communication. Operating on a broader spectrum of frequencies with greater flexibility and power, ham radios require a license but enable communication with operators globally. In emergencies when traditional methods fail, ham radios become invaluable tools.

Now, the crucial question: how do we select the right handheld radio? Consider factors such as range, battery life, and features. For basic short-range communication, an FRS radio suffices. However, for ventures off the beaten path or the need for longer distances, a GMRS or ham radio offers superior range and versatility.

Effectively operating handheld radios in off-grid scenarios requires practice. Familiarize yourself with the controls and functions of your chosen device. Experiment with different channels and frequencies to determine optimal communication routes for your location. Investing in rechargeable batteries and a solar charger ensures your radio remains powered when the grid fails.

So, handheld radios stand as the backbone of off-grid communication. Whether traversing mountains or preparing for disaster, having a reliable handheld radio in your arsenal can mean the difference between staying connected and being stranded in the dark.

In the wild and untamed expanses, where cell towers dare not tread and Wi-Fi signals faint into oblivion, there lies a realm where traditional communication methods falter. But fear not, fellow survivalist, for in this modern age, satellite communication devices emerge as our lifelines in remote areas.

Let's dive straight into the heart of the matter: understanding satellite communication technology. Picture this – satellites orbiting high above the Earth, tirelessly relaying our messages across vast distances. These marvels of modern engineering ensure that no corner of the globe remains beyond our reach. Unlike traditional cell towers that rely on terrestrial infrastructure, satellite devices bypass these limitations, offering connectivity where others dare not venture.

Now, onto the gear. Satellite phones, messengers, and trackers – the triumvirate of communication prowess in the wilderness. Satellite phones, akin to their terrestrial counterparts, allow for voice calls and text messaging even in the most remote locales. Imagine standing atop a rugged peak, with only the vast expanse of nature for company, yet with a simple push of a button, you're connected to the outside world.

But wait, there's more. Satellite messengers step up the game, offering not just communication, but also the ability to send SOS signals in times of dire need. These compact devices, often no larger than a smartphone, pack a punch when it comes to keeping you safe and connected. Picture this scenario: You're trekking through dense forests, miles away from civilization, when disaster strikes. With a satellite messenger in hand, help is just a distress signal away.

And let's not forget about trackers – the silent guardians of the wilderness. These devices allow your loved ones back home to keep tabs on your whereabouts, offering peace of mind and reassurance with every step you take. Whether you're traversing icy tundras or navigating sweltering deserts, a satellite tracker ensures that your journey is never truly solitary.

Now, onto the nitty-gritty – subscription plans and coverage areas. Before you embark on your wilderness odyssey, it's crucial to research and select the right plan for your needs. From pay-as-you-go options to monthly subscriptions, there's a plan tailored to suit every adventurer's budget and requirements. And when it comes to coverage, rest assured that satellite communication services span the globe, ensuring that even the most remote corners remain within reach.

Let's paint a picture with a real-life case study. Meet Tara, an intrepid explorer with a passion for adventure. Setting her sights on the rugged peaks of the Himalayas, she knows that communication is key to her survival. Armed with a satellite phone and messenger, she embarks on her journey, traversing treacherous terrain and facing the elements head-on. When a sudden blizzard leaves her stranded at high altitude, Tara activates her satellite messenger, alerting rescue teams to her predicament. Thanks to her trusty devices and quick thinking, Tara emerges from her ordeal unscathed, a testament to the power of satellite communication in the wild.

In the realm of off-grid communication, where traditional methods falter and modern infrastructure fades into oblivion, there exists a beacon of hope: mesh networking devices. These ingenious gadgets are the cornerstone of building local communication networks that defy the constraints of distance and terrain.

Picture this – a web of interconnected devices, each acting as a node in a sprawling network. Unlike traditional communication systems that rely on centralized hubs, mesh networks are decentralized, empowering each node to communicate directly with its neighbors. This redundancy ensures that even if one node fails, the network remains resilient, steadfast in its mission to keep you connected.

Now, onto the gear. Portable Wi-Fi routers and mesh radio systems – the stalwarts of mesh networking in the wild. Portable Wi-Fi routers, compact and versatile, transform any location into a hub of connectivity. With the ability to create ad-hoc networks on the fly, these devices are indispensable for establishing communication hubs in remote areas. Imagine setting up camp in the heart of the wilderness, surrounded by towering trees and rugged terrain, yet with a portable Wi-Fi router at your side, you're never truly isolated.

But wait, there's more. Mesh radio systems take communication to the next level, offering long-range connectivity without the need for traditional infrastructure. These robust devices, often ruggedized for outdoor use, allow for voice communication over vast distances, bridging the gap between remote outposts and urban centers. Picture this scenario: You're part of a search and rescue team, scouring the wilderness for a missing hiker. With a mesh radio system in hand, you coordinate with your team members, ensuring that every step brings you closer to your goal.

Now, onto the nitty-gritty – setting up and configuring mesh networks. Before you start on your off-grid adventure, it's crucial to familiarize yourself with the intricacies of mesh networking. From choosing the right devices to optimizing network topology, there are several factors to consider when building your local communication network. Fear not, dear reader, for with a bit of know-how and perseverance, you'll soon have a robust mesh network at your fingertips.

Let's paint a picture with a real-life case study. Meet Jack, a seasoned outdoorsman with a passion for exploration. Setting his sights on the remote wilderness of Alaska, Jack knows that communication is key to his survival. Armed with a portable Wi-Fi router and mesh radio system, he embarks on his journey, traversing icy tundras and snow-capped peaks. When a sudden blizzard strikes, threatening to engulf him in its icy embrace, Jack activates his mesh network, rallying his companions and ensuring that no one is left behind. Thanks to his foresight and preparedness, Jack emerges from the storm unscathed, a testament to the power of mesh networking in the wild.

In the world of off-grid communication, where the grid's comforting hum fades into silence, the need for reliable power sources is paramount. Picture this: you're deep in the wilderness, your trusty gadgets by your side, but their batteries are dwindling, and there's no outlet in sight. Fear not, my fellow survivalist, for with solar-powered charging solutions, you can keep your devices powered up, no matter how far off the beaten path you roam.

First and foremost, let's talk about the importance of reliable power sources for off-grid communication devices. In a world where connectivity is king, a dead battery can mean the difference between safety and peril. Whether it's a smartphone for emergency calls, a radio for communication with fellow adventurers, or a GPS device for navigation, keeping these gadgets powered up is essential for your well-being in remote areas.

Now, onto the gear. Solar charging options come in all shapes and sizes, catering to the diverse needs of off-grid explorers. From compact solar panels that fit in your backpack to rugged solar chargers designed to withstand the harshest conditions, there's a solution for every situation. Picture this scenario: You're trekking through the desert, the sun beating down relentlessly, when your radio's battery starts to fade. With a solar charger strapped to your pack, you harness the power of the sun to keep your device juiced up and ready for action.

But wait, there's more. Maximizing solar charging efficiency is key to keeping your gadgets powered up when off-grid. Here are a few tips to help you get the most out of your solar charging set-up:

- Position your solar panels to capture the most sunlight throughout the day.
- Keep your devices shaded and cool while charging to prevent overheating.
- Invest in high-quality solar panels and chargers to ensure reliability in challenging conditions.
- Use energy-efficient gadgets and prioritize charging based on necessity to conserve power.

Solar-powered charging solutions are more than just gadgets – they're lifelines in remote areas, keeping your devices powered up when the grid fades away. So before you venture off-grid, equip yourself with the tools that will keep you connected and safe, no matter where your adventures may take you.

When you find yourself at the world's end, where the vast expanse of wilderness stretches out before you with no civilization in sight, your ability to signal for help becomes paramount. In the realm of wilderness survival, mastering signaling techniques can mean the difference between rescue and isolation.

Signalling Techniques.

Let's start with the basics: an overview of signaling techniques using visual, auditory, and tactile methods. Visual signaling involves anything that can catch the eye of a passing aircraft or search party. This could be as simple as waving brightly colored clothing, or as sophisticated as using signal mirrors or smoke signals. Auditory signaling, on the other hand, relies on sound to attract attention. This can be achieved through shouting, blowing a whistle, or banging on objects to create rhythmic patterns. Lastly, tactile signaling involves creating physical markers or leaving clues for rescuers to follow, such as marking trails with symbols or leaving behind notes.

Signal mirrors, often overlooked but incredibly effective, harness the power of sunlight to create a flash that can be seen from miles away. When using a signal mirror, hold it at arm's length and angle it towards the sun, then aim the reflection towards your intended target. With a bit of practice, you can create a bright flash that is sure to catch the attention of anyone nearby.

Next up, whistles – small but mighty tools that can penetrate even the densest forests. When using a whistle for signaling, three short blasts followed by a pause is the universally recognized distress signal. Keep your whistle handy and practice blowing it regularly to ensure that you can produce a loud, clear sound when it counts.

And let's not forget about signal fires – the time-honored method of signaling for help in the wilderness. When building a signal fire, aim for maximum visibility by choosing an open area with good airflow. Use dry tinder and kindling to get the fire started, then add green vegetation to create thick smoke that will billow into the sky. Remember to keep the fire contained and under control to avoid inadvertently causing a wildfire.

Case study: Meet Tom, an experienced outdoorsman with a passion for adventure. While hiking in the remote mountains of Colorado, Tom finds himself injured and unable to continue. With no cell service and no one in sight, he turns to his signaling skills to attract help. Using his signal mirror, Tom catches the attention of a passing helicopter, which spots his location and airlifts him to safety. Thanks to his quick thinking and mastery of signaling techniques, Tom is reunited with his loved ones, a testament to the power of creativity and resourcefulness in the wild.

So, when you find yourself at the world's end, remember that help may be closer than you think. By mastering signaling techniques and staying calm under pressure, you can increase your chances of being found and rescued in even the most remote wilderness settings. So before you embark on your next adventure, take the time to learn and practice these essential skills – they just might save your life.

Morse Code

In the annals of communication history, there exists a language both elegant and timeless – Morse code. Developed in the early 19th century by Samuel Morse and Alfred Vail, Morse code revolutionized long-distance communication, serving as the backbone of telegraphy for decades. Today, in our modern age of instant messaging and digital connectivity, Morse code may seem antiquated, but make no mistake – its relevance endures, especially in emergency situations where traditional communication methods falter.

Morse code is a language of dots and dashes that transcends barriers of distance and technology. In Morse code, each letter of the alphabet, as well as numbers and punctuation marks, is represented by a unique combination of short and long signals. These signals, when transmitted and decoded, convey messages with remarkable efficiency and clarity.

Now, onto the methods of transmitting Morse code. When it comes to off-grid communication, creativity is key. Visual signaling offers a straightforward way to transmit Morse code using light signals. Imagine flashing a flashlight in the darkness, spelling out S-O-S in Morse code – three short flashes, three long flashes, three short flashes. With a bit of practice, you can communicate vital information over long distances using nothing but light and shadow.

But what if you find yourself in a situation where visual signaling is not feasible? Fear not, my friend, for sound signals offer an alternative method of transmitting Morse code. Whether it's tapping on a pipe, blowing a whistle, or banging on objects, the rhythmic pattern of short and long signals can cut through the silence and reach those in need. Picture this scenario: You're lost in the wilderness, far from civilization, when you hear the faint sound of someone tapping out Morse code on a tree trunk. With a keen ear and knowledge of Morse code, you decipher the message and signal back for help.

And let's not overlook tactile signaling – a method of transmitting Morse code using touch rather than sight or sound. Whether it's tapping on someone's shoulder or using hand signals to convey messages, tactile communication can be a lifeline in situations where other methods fail. Imagine being trapped in a collapsed building, unable to see or hear your rescuers, but able to feel their touch as they tap out Morse code on your hand.

Morse code is more than just a relic of the past – it's a versatile and reliable communication tool that has stood the test of time. So before you venture off-grid, take the time to learn and practice Morse code – it just might save your life when other methods fail.

Building Emergency Beacons and Markers

When it comes to emergency signaling, tapping into the resources nature provides is essential. In the wild, where human-made structures are scarce, making the most of natural features becomes crucial. Let's explore how you can harness the environment around you for effective signaling, increasing your chances of being found in an emergency.

To start, let's emphasize the significance of utilizing natural features such as open areas, water bodies, and elevated terrain for signaling. These elements offer visibility and contrast against the surrounding landscape, serving as ideal platforms for sending distress signals. Whether it's a clearing in the forest, a glistening lake, or a lofty hilltop, these natural features act as beacons of hope in desperate times.

There are various techniques for creating ground-to-air signals and enhancing visibility using natural materials. One straightforward method involves arranging rocks, branches, or other debris to spell out letters or symbols easily recognizable from the air. Imagine arranging stones to form the letters S-O-S on a beach or using branches to create an arrow pointing towards your location. Such signals catch the attention of passing aircraft or search parties, guiding them straight to you.

But what if you're in an area devoid of human-made materials? Fear not, for nature provides an abundance of resources for signaling. Take, for instance, the humble smoke signal. By kindling a fire using damp vegetation or greenwood, you can produce thick, billowing smoke that rises into the sky, visible from great distances. Select an elevated location with ample airflow and add green foliage or other materials to generate smoke that stands out against the backdrop.

Water bodies also offer unique opportunities for signaling. By employing reflective materials or creating patterns on the water's surface, you can attract attention from passing boats or aircraft. Picture using a mirror to catch the sunlight and create flashes that dance across the waves or arranging brightly colored objects on the shoreline to form a visible pattern.

In the wild, when faced with an emergency, remember that nature is your ally. By making use of natural features and materials for signaling, you can enhance your visibility and improve your chances of being found and rescued. So, before embarking on your next adventure, familiarize yourself with these techniques—they just might save your life when you need them most.

As we draw the curtains on our exploration of emergency preparedness and off-grid communication, it's time to take stock of the practical skills and knowledge we've amassed. From mastering the ins and outs of radio communication to delving into creative methods of signaling for help, each chapter has been a roadmap for self-reliance in times of crisis.

Throughout our journey, we've peeled back the layers of complexity surrounding radio communication, from understanding frequency bands to selecting the right handheld radio for our needs. We've uncovered the lifelines provided by satellite communication devices and mesh networking technology, ensuring we're never truly cut off from the world, even when phones fail.

Morse code and signaling techniques have become second nature, offering us a timeless method of communication when other means falter. And with a keen eye for utilizing natural features and constructing improvised beacons, we've equipped ourselves with the tools to attract attention and aid in the most remote of locations.

But let's not forget the essence of preparedness – planning and protocol. By developing communication plans and assigning roles within our groups, we've laid the groundwork for effective coordination when it matters most. And by implementing encryption measures and verification protocols, we've ensured the security and reliability of our communications in the face of adversity.

Our journey has been grounded in real-world applications, from the stranded boater using a smoke signal to the lost camper marking a trail with reflective tape. These case studies serve as reminders of the tangible impact our knowledge and actions can have in moments of crisis.

So as we part ways, let us carry forward the lessons learned and the skills acquired. Let us embrace the pragmatic truth that preparation is the key to survival in uncertain times. And let us stand ready, armed with the tools and know-how to navigate the challenges of emergencies with practicality, determination, and a steadfast commitment to self-reliance.

Book 7: Ham Radio – The Voice of Hope

Welcome to a realm where readiness is not just planned but enacted, where the ability to communicate transcends mere convenience and becomes essential for survival. This is not merely another manual; it's your entryway into the indispensable field of ham radio, tailored for those who understand that in the aftermath of disruption, the ability to reach out, to connect, and to coordinate is not just advantageous, it's vital.

You're starting on this journey because you recognize the importance of being one step ahead, of mastering the art of communication when the usual channels are obliterated. You are here to grasp the robust potential of ham radio – an instrument that becomes your bridge to the outside world when all else falls silent.

As you navigate through these pages, you'll uncover the essentials of establishing and maintaining your emergency communication channel. You'll be equipped not just with theoretical insights but with practical strategies, ensuring that when the final page turns, you're not merely more informed but unequivocally more prepared.

In the world of emergency preparedness, communication is king. This is where ham radio, an often underestimated but critically important tool, comes into play. It's not just an old-school hobby; it's a backbone of reliable, independent communication in crises. Let me walk you through why mastering ham radio should be at the top of your list if you're serious about self-reliance.

A ham radio isn't just any ordinary radio. Picture this: while typical radios in your car or on your nightstand only let you listen in, a ham radio lets you talk back, turning one-way chatter into a two-way conversation. What sets it apart from your everyday radio and other modern communication gadgets? It's all about the self-reliance and the skills. With a ham radio, you're not just a passive receiver; you're an active participant. You're not dependent on cell towers, internet connections, or commercial networks that might buckle when disaster strikes. Instead, you've got a tool that works under almost any condition, anywhere. But here's the kicker: to get on these airwaves, you need to earn your stripes through licensing, proving you've got the chops to operate safely and effectively. That's ham radio – it's hands-on, it's independent, and it's a lifeline when all else fails.

let's paint a picture of what a ham radio setup might look like – this isn't your average piece of tech. Picture a robust, sturdy piece of equipment, one that feels like it's got a story to tell. It's not sleek or flashy; it's functional, built to last, and looks like it means business.

Your typical ham radio setup might include a transceiver – that's the heart of the operation, where the magic of sending and receiving signals happens. It's usually a boxy, metal unit, filled with knobs, buttons, and dials, each serving a unique purpose, from tuning into the right frequency to adjusting the volume. The front panel's often lit up with displays or meters, giving you real-time feedback on signal strength and other vital stats.

Then there's the microphone, which might be a handheld piece or a headset, depending on your setup and style. It's your direct line, turning your words into signals that travel miles, sometimes even bouncing off the ionosphere to reach distant lands.

And we can't forget about the antenna – the unsung hero of the ham radio world. It could be a simple wire strung up between trees, a complex beam mounted on a tall mast, or even a mobile antenna attached to your vehicle. The antenna is what sends your voice soaring over the hills, across the seas, or into the sky, turning your local setup into a global communication station.

Let's start with a brief history to set the stage. Ham radio, or amateur radio, isn't a product of the modern era's digital revolution. It's been around since the late 1800s, evolving alongside advancements in technology. Initially, it was a way for enthusiasts to experiment and communicate, but its true potential shone through during times of emergency. The history of ham radio is as fascinating as it is diverse, stretching back over a century, intertwining with the very fabric of modern technological advancements.

The story begins in the late 19th century, in the pioneering days of wireless communication. Figures like Nikola Tesla and Guglielmo Marconi were laying the groundwork for wireless telegraphy, unknowingly setting the stage for what would become ham radio. Initially, these wireless technologies were the realm of scientists and large commercial enterprises. However, as the technology evolved, it captured the imagination of enthusiasts and hobbyists, giving birth to the amateur radio movement.

During the early 20th century, amateur radio operators began to populate the airwaves, experimenting with equipment, antennas, and frequencies. These early hams were pioneers, often building their radios from scratch, pushing the boundaries of what was possible with wireless communication. But it wasn't just about tinkering with technology; amateur radio quickly proved its value to society, especially in times of crisis.

One of the most notable early uses of ham radio was during the Titanic disaster in 1912, where wireless communication played a crucial role in the rescue operations. This event highlighted the potential of wireless communication for emergency response, a theme that has remained central to ham radio culture.

The world wars further underscored this, as amateur radio operators provided essential communication services, bridging gaps left by destroyed infrastructure. Post-war periods saw a boom in ham radio, with veterans and civilians alike joining the ranks, spurred by advancements in technology and a growing sense of global community.

The latter half of the 20th century and the onset of the 21st brought about significant technological advancements, from transistors to digital communication, each revolutionizing ham radio in different ways. Today, ham radio is a blend of the old and the new, combining the traditional aspects of radio communication with cutting-edge technologies like digital modes and satellite communication. Over the years, ham radio operators have been instrumental in providing critical communication lines during natural disasters, wars, and other crises when traditional channels fail.

While ham radio is renowned for its role in emergency communication, its uses extend far beyond the confines of disaster response. Amateur radio has found its way into a myriad of unconventional and fascinating areas, showcasing its versatility and adaptability.

One such area is scientific research. Ham radio operators have contributed to scientific endeavors, such as tracking migratory patterns of endangered species, monitoring environmental conditions, and even bouncing radio signals off the moon to study celestial phenomena—a practice known as Earth-Moon-Earth (EME) communication.

Space communication is another frontier where ham radio has made its mark. The Amateur Radio on the International Space Station (ARISS) program allows students and enthusiasts to communicate directly with astronauts, providing a unique educational experience and fostering international goodwill. Furthermore, ham radio satellites, known as CubeSats, launched by amateur radio organizations, orbit the Earth, enabling operators worldwide to experiment with space communication.

In remote and underdeveloped areas, ham radio serves as a bridge to the outside world, offering educational services, health information, and a means of connecting communities otherwise cut off from mainstream communication networks. This has proven particularly valuable in regions where internet access is nonexistent or prohibitively expensive.

Culturally, ham radio has transcended communication, fostering international friendship and understanding. Special event stations, DXpeditions (long-distance ham radio expeditions), and contests bring together operators from different corners of the globe, celebrating cultural events, historical milestones, and shared interests.

Ham radio has also found a place in art and entertainment, with operators integrating Morse code into music, utilizing radio frequencies in sound art installations, and even transmitting digital images across the globe. These creative applications highlight the blend of technology and artistry inherent in ham radio, expanding its appeal beyond the traditional tech-savvy circles.

Now, you might wonder why ham radio is so crucial for anyone focused on self-reliance. The answer is simple: independence and reliability. In the face of a disaster, conventional communication tools like phones and the internet can become useless, severed by the very forces wreaking havoc. Ham radio, on the other hand, operates independently of these systems. It doesn't rely on cell towers, satellites, or internet connections. Instead, it uses radio waves to cut through the chaos, connecting people when it matters most.

This brings us to the pivotal role of amateur radio in emergencies. Imagine a scenario where a natural disaster has knocked out all forms of communication. The local infrastructure is shattered, and with it, the community's lifeline to the outside world. Here, ham radio operators step in, transforming their homes into makeshift communication hubs. They become the critical link between the affected community and emergency services, coordinating rescue and relief efforts effectively.

But it's not just about having the equipment; it's about knowing how to use it. This is where the practical, hands-on approach comes into play. If you're aiming to be genuinely self-reliant, it's not enough to own a ham radio set; you must be proficient in its use. This means regular practice, understanding the nuances of radio frequencies, and being able to construct or repair equipment as needed.

Consider a real-life case where this knowledge made a tangible difference. After a devastating hurricane, a small coastal town was completely cut off. The local ham radio operator, a regular guy with a passion for preparedness, managed to establish communication with emergency responders. His actions facilitated the delivery of medical supplies and guided rescue teams to the most affected areas. His knowledge and readiness to act turned him from a quiet neighbor into a community hero.

This story illustrates not just the utility of ham radio in emergencies but the power of individual preparedness. As we look more deeply into the world of ham radio, remember, this is more than a tool; it's a lifeline, a means to ensure that when the world goes silent, your voice can still be heard. It's about taking control, stepping up, and being ready to make a difference. In the landscape of emergency preparedness, ham radio stands out as a beacon of hope, a tool that empowers you to stay connected, informed, and, above all, safe.

here are the practical steps and tips to help you navigate the process of obtaining a ham radio license and becoming an active part of the ham radio community:

15. Start with Research: Begin by understanding the different classes of ham radio licenses available and determine which level suits your needs and interests. The three main types in many countries are Technician, General, and Extra, each offering different privileges.
16. Gather Study Materials: Collect the necessary study materials for your chosen license level. Resources can include books, online courses, and practice tests. The ARRL (American Radio Relay League) website is a great place to start for reliable study guides and materials.
17. Join a Study Group: Consider joining a local or online study group. Learning with others can provide additional insights, motivation, and clarification of complex topics.
18. Set a Study Schedule: Dedicate regular time slots each week for study. Breaking down the material into manageable sections can make the process less overwhelming and more effective.
19. Take Practice Exams: Regularly complete practice exams to gauge your understanding and

readiness for the actual test. Many websites offer free practice tests that simulate the format of the official exam.

20. Find a Test Session: Once you feel prepared, locate a nearby test session. The ARRL website, as well as local ham radio clubs, can provide information on test dates and locations.
21. Apply for the Exam: Contact the test session administrator to register for the exam. Make sure you understand the requirements, including any identification or fees needed.
22. Attend the Test: On the day of the exam, bring the necessary identification and materials as instructed. Arrive early to fill out any paperwork and to settle in.
23. Engage with the Community: After passing the exam and receiving your call sign, start engaging with the ham radio community. Attend club meetings, participate in local nets, and practice making contacts.

By following these steps, you'll not only gain your ham radio license but also start on a path to becoming an integral part of a global community dedicated to communication, support, and emergency preparedness.

When it comes to setting up your emergency network through ham radio, the right approach is clear-cut: choose wisely and set up efficiently. This isn't about having the most expensive or the most advanced equipment; it's about having the right tools for the job that match your budget, your space, and your specific needs.

Firstly, selecting the right equipment can seem daunting with all the options available, but it boils down to understanding your requirements and constraints. Start with the basics: a reliable transceiver, a power source, and an antenna. Your choice of transceiver should align with your license level and intended use—whether that's local communication on VHF/UHF bands or long-distance on HF bands. If you're starting out, consider a dual-band VHF/UHF handheld or mobile transceiver; they're more affordable and cover many emergency communication needs.

Your budget is crucial here. Don't stretch yourself thin trying to buy top-of-the-line gear all at once. Begin with essential, good-quality equipment, and expand your setup as you gain experience and understand your needs better. Also, consider the second-hand market, where you can find excellent equipment at a fraction of the cost—just ensure it's from a reputable source and in good working condition.

Space is another factor. If you're limited, a handheld may be your best start, coupled with a simple, effective antenna like a magnetic mount or a small Yagi. If you have more room, you might opt for a base station setup with a larger antenna, offering you greater range and versatility.

Essential Ham Radio Equipment Checklist:

1. Transceiver: Choose between handheld, mobile, or base station models based on your communication needs and space.

2. Power Source: Invest in a reliable power supply, battery pack, or solar charger.
3. Antenna: Select based on your location's space and the frequencies you plan to use (e.g., a dual-band antenna for VHF/UHF).
4. Coaxial Cable: Quality matters here for connecting your antenna to your transceiver with minimal signal loss.
5. SWR (Standing Wave Ratio) Meter: Essential for tuning your antenna and minimizing transmission issues.
6. Headphones or Speaker: For clearer audio reception in noisy environments.
7. Microphone or Handset: Depending on your transceiver's requirements and your personal preference.
8. Grounding Equipment: Including grounding rod, wire, and lightning arrester for safety and equipment protection.

Now, setting up your station is where the rubber meets the road. The location of your antenna is critical; it should be as high and as clear of obstructions as possible to maximize range and effectiveness. Whether you're installing a simple dipole, a vertical, or a more complex antenna system, ensure it's well-supported and securely mounted. Safety is paramount—keep antennas away from power lines and areas of high foot traffic.

Grounding your equipment is not just a safety measure; it's essential for operational efficiency and protecting your gear from electrical surges. Make sure your station has a good electrical ground and consider a lightning arrester for added protection, especially if your antenna is outdoors and high up.

A real-world application of this came during a severe storm in a rural area. A local ham, Lisa, had set up her station according to these principles. Despite power outages and the collapse of cell networks, her station remained operational. Using her battery-backed setup and a well-placed wire antenna, she managed to establish communication with emergency services and coordinate aid for her isolated community. Her preparation and understanding of effective station setup proved invaluable, not just for her own safety, but for her entire community.

Remember, your ham radio station is more than equipment and antennas; it's your lifeline during emergencies and your bridge to the wider community. Invest the time to set it up correctly, practice using it, and you'll be well-prepared to provide aid when the unexpected occurs. This isn't just about being ready; it's about being effective when it matters most.

Key Maintenance Tasks for Your Ham Radio Equipment:

- Cable and Connector Checks: Inspect for any signs of wear, frays, or corrosion.
- Battery Maintenance: Charge regularly, store in a cool, dry place, and rotate usage.
- Antenna Inspection: Look for damage or wear and ensure it is securely mounted and free from obstructions.

- Clean Equipment: Dust off all surfaces and keep the equipment dry to prevent damage.
- Software Updates: Regularly update any software associated with digital ham radios to ensure optimal performance.

In the realm of emergency preparedness, your equipment isn't just gear—it's your lifeline. That's why maintaining your gear with meticulous care isn't optional; it's essential. You wouldn't neglect your vehicle's maintenance, letting it fall into disrepair, because you count on it. Apply the same principle to your ham radio equipment. Your life and the lives of others could depend on its reliability in a crisis.

Start with regular inspections of your equipment. Check cables for frays, connectors for rust or corrosion, and all equipment for dust and moisture build-up. These are your basics, but they're crucial. A malfunction due to neglected maintenance can mean the difference between being a lifeline and being left in silence when disaster strikes.

Battery care is paramount. Regularly check the charge, ensure they're stored in a cool, dry place, and cycle through your stock, so they're used evenly. For rechargeable types, maintain them according to manufacturer instructions to maximize lifespan. Remember, a powerless radio is as good as a brick in an emergency.

Antenna maintenance is equally vital. Inspect it for any signs of wear or damage, especially after severe weather. Ensure it's securely mounted and that the surrounding area remains clear of obstructions that could impede signal transmission or reception.

But what's good gear without the know-how to use it? This is where the mantra "Practice makes perfect" takes center stage. Regular use of your ham radio isn't just about staying familiar with its functions; it's about ensuring it remains in good working order and improving your skills over time.

Make sure to participate in local nets. These are not just opportunities to socialize; they're invaluable practice sessions. You'll learn the local lingo, understand how to break through interference, and get comfortable with protocols. This isn't just about chatting; it's about honing your ability to communicate effectively under various conditions.

But practice isn't just about talking; it's also about listening. Tune into different frequencies, listen to other operators, and get a feel for how they handle communications. The more you listen, the more you learn—about handling emergencies, about etiquette, and about the vast world of ham radio.

Now, consider the case of John, a ham operator who had maintained his equipment with routine checks every few months. He lived in a region prone to wildfires and, unfortunately, one summer, his area was hit hard. But John was prepared. His equipment was in top condition, his batteries were fully charged, and, thanks to his regular participation in local nets, he was well-practiced in emergency communication.

When the fires disrupted local cell and internet services, John's radio became the primary link between his neighborhood and the outside world. He coordinated with emergency services, relayed critical information to his neighbors, and provided a calm, knowledgeable presence in the midst of chaos. His preparation and practice made all the difference.

In the world of emergency preparedness, isolation is your enemy, and connection is your ally. It's not just about having the right gear and know-how; it's about building a network, both locally and globally, before you ever need to send out a distress call. This is where ham radio comes into its own, not just as a tool but as a bridge between you and the rest of the preparedness community.

Let's get down to brass tacks. Building your network with ham radio isn't about casual chit-chat; it's about establishing reliable, strong connections that could one day be a lifeline. Start locally. Reach out to local ham radio clubs or emergency response groups. These are your immediate go-to people when things go south. Engage with them, learn from them, and become a regular voice on local nets. This isn't just networking; it's building a mutual support system.

But don't stop at your local community. Ham radio gives you the incredible advantage of global reach. Start making regular contacts in wider areas, join international nets, and participate in global events like Field Day or Jamboree-on-the-Air. Each contact you make, each relationship you build, is a potential resource in a crisis.

Remember, the key here is consistency. Make regular contacts, be reliable, and contribute positively to your networks. When you reach out, it's not just about making noise; it's about offering value, whether it's technical advice, local weather reports, or just a steady presence. Over time, this builds trust and reliability – essential components of a strong emergency network.

Now, let's pivot to the crucial aspect of emergency communication protocols. In a crisis, clarity saves lives. Knowing the right protocols and etiquettes is non-negotiable. First and foremost, understand the concept of 'listening first.' In emergencies, the airwaves can be chaotic. Tune in and listen before you transmit to ensure you're not interrupting critical communications.

Next, familiarize yourself with standard emergency communication formats, such as the use of phonetic alphabet. This isn't just military jargon; it's a universal tool to avoid misunderstanding. Say your call sign is K7XYZ. In a stressed situation, static or interference could make that hard to understand. So, you'd say, "Kilo Seven X-ray Yankee Zulu." Each word stands out clearly, reducing the chance of miscommunication. Imagine you need to report a license plate, location, or critical detail; using the phonetic alphabet can be the difference between being understood and being misheard, which, in emergencies, could mean the difference between help arriving or not.

The Incident Command System (ICS) is equally as important. Imagine there's a fast-moving brush fire near your community. As a licensed ham radio operator, you're part of the emergency communication chain. Using ICS principles, you'd first identify yourself and your location using clear, concise language. You'd report: "This is [Your Call Sign], operating at the northern edge of Pineview Community, observing southward advancing brush fire, over."

In ICS, you're a piece of a larger puzzle, fitting into a structured response system. You would report specifics: the fire's size, direction, speed, and proximity to homes. If you're in direct contact with local emergency services, you'd use ICS terminology they recognize — reporting your situation as "immediate," "urgent," or "routine" based on the threat level. By sticking to this system, you help streamline the flow of information, enabling a coordinated response without causing additional confusion or panic.

This isn't bureaucracy; it's about ensuring your message is understood clearly and immediately by any responder. These tools are your allies in ensuring that when you reach out for help or to provide information, you're understood, efficient, and effective. Integrating these into your ham radio practices makes you not just a communicator, but a lifeline when every second counts.

Moreover, learn to be concise. In emergencies, time is precious. Relay only vital information: Who you are, your location, the nature of the emergency, and the type of assistance needed. Stick to facts, avoid speculation, and always follow the directions of emergency operators or net controllers.

In the rugged landscape of emergency preparedness, where uncertainty is the only certainty, ham radio emerges not merely as a piece of equipment, but as a beacon of hope. Take, for instance, the relentless hurricanes that pummel coastlines. In one harrowing episode, when traditional communication systems were decimated, it was the ham radio operators who rose to the occasion. They didn't just communicate; they coordinated rescue operations, relayed messages from worried relatives, and provided real-time updates to emergency services. This wasn't theatre; it was reality, with lives hanging in the balance, showcasing the undeniable value of ham radio under pressure.

But the value of ham radio extends beyond natural disasters. In the chaos following a massive urban blackout, where darkness swallowed the city and silence reigned, a network of ham operators became the voice in the darkness. They turned their homes into makeshift command centers, guiding panicked citizens and facilitating communication between emergency personnel. This scenario wasn't a test. It was a stark reminder that when modern conveniences fail, the age-old ham radio stands strong.

These stories are not just tales of heroism; they are blueprints for preparedness, emphasizing the critical need for continuous education and engagement within the ham radio community.

Continuing education in ham radio is not a mere suggestion; it is a necessity. As operators, the responsibility to maintain a cutting edge understanding of radio operations and emergency protocols lies heavily upon us. The landscape of radio technology is ever-changing, with new advancements and methodologies continually emerging. Diving back into the books, attending advanced licensing classes, or exploring new digital modes of communication can significantly elevate your operational capabilities. Remember, in the midst of a crisis, your knowledge can spell the difference between chaos and order.

However, knowledge without application is like a radio without power. This is where the invaluable role of community engagement comes into play. Immersing yourself in the local and global ham radio community isn't just about expanding your social circle; it's a strategic move to enhance your preparedness. Active participation in field days, engagement in weekly nets, and collaboration with emergency response teams provide practical experiences that are as crucial as any theoretical knowledge. These interactions forge bonds, build trust, and create a robust network of individuals who are not only familiar with each other's voice but are also well-versed in each other's capabilities and operational style.

Moreover, volunteering for emergency drills with local agencies or offering your expertise in community preparedness seminars not only solidifies your standing as a reliable operator but also deepens your understanding of the real-world application of ham radio in crisis situations. This continuous loop of learning, applying, and sharing not only hones your skills but also reinforces the collective strength of the ham radio community.

In the grand scheme of emergency preparedness, where unpredictability is the norm, remaining static is not an option. The stories of past emergencies serve as stark reminders of the critical role ham radio plays. Therefore, take the initiative: dive deeper into your education, weave yourself into the fabric of the ham radio community, and stand ready. When the unforeseen strikes, be more than just prepared—be indispensable. Your actions today define the outcome of tomorrow's emergencies. Let's not just be ready; let's be poised for action, with our radios charged, our skills sharp, and our networks strong.

As we close the final pages of this guide, it's clear that the journey into the world of ham radio and emergency preparedness is far from over. In fact, consider this ending merely the first checkpoint in a much longer, critical path towards self-reliance and robust preparedness. This book has been more than just words on a page; it's a call to action—a toolkit for those ready to take responsibility for their own safety and the well-being of those around them.

Remember, the essence of this guide isn't rooted in theoretical knowledge alone but in the actionable steps and real-world applications you've been equipped with. You now have the foundation to not just survive, but to thrive in the face of adversity. The real-life scenarios and case studies highlighted throughout were not just stories; they were lessons carved out of the realities of emergency situations, proving time and again the indispensable value of ham radio communication.

But knowledge without action is like a radio without power. It's imperative that you don't let this newfound understanding sit idle. Test your equipment regularly, engage with your local ham radio community, and keep pushing the boundaries of your skills and knowledge. Emergencies won't wait for you to be ready; you need to maintain a state of constant preparedness, always looking ahead, always learning.

Moreover, the responsibility doesn't end with you. Share your knowledge, encourage others to join the ham radio community, and build a network of prepared individuals. The stronger your community, the more resilient you all become. In times of crisis, these connections can mean the difference between isolation and support, between uncertainty and coordinated action.

The path ahead is yours to shape. Armed with your ham radio, bolstered by knowledge, and supported by a community of like-minded individuals, you stand ready to face whatever challenges may come. This isn't just preparation; it's empowerment. So, take the next step, key up your radio, and step confidently into the world of emergency preparedness. The airwaves await, and so does your journey into self-reliance and proactive readiness.

Book 8: Satellites and Digital Tech: The Future of Survival

Gazing skyward isn't just for dreamers. In our world, it's a pragmatic approach to safety and communication amid disasters. Satellites, high above Earth's turmoil, offer a steady and wide-reaching connection. Satcom, the tech behind this, bridges vast spaces in seconds. Far from being sci-fi, this is essential gear for those serious about survival.

Why trust satellites? Earth-bound communication networks fail under nature's wrath. Hurricanes, earthquakes, or fires can cripple cell towers and landlines. But satellites, in their high orbit, remain untouched, ever-ready to transmit distress signals, coordinate aid, and guide the lost.

Satellite phones are used in a variety of situations where traditional cell phone coverage is unavailable or unreliable. Their primary function is to provide voice communication, similar to regular mobile phones, but they can also send and receive text messages and, in some cases, offer data services for email and basic internet access. Here are some common scenarios and uses for satellite phones:

- Emergency Response and Disaster Relief: In the aftermath of natural disasters like hurricanes, earthquakes, or floods, when local communication infrastructure may be damaged or overwhelmed, satellite phones become crucial.
- Adventure and Exploration: Adventurers, explorers, and researchers working in remote areas of the world, such as deep-sea explorers, polar scientists, and jungle expedition teams, rely on satellite phones to stay in touch with the outside world, share their findings, and call for assistance in emergencies.
- Maritime Communication: Crew members on ships and ocean-going vessels use satellite phones as a primary means of communication when out of range of land-based cellular networks.
- Aviation: In remote areas, pilots of small aircraft may carry satellite phones as a backup communication tool, enabling them to contact air traffic control or emergency services if their main communication systems fail.
- Remote Work Locations: Workers in the mining, oil, and gas industries often find themselves in isolated areas where conventional communication systems are not available.
- International Travel and Journalism: Travelers, journalists, and correspondents working in conflict zones or countries with restrictive telecommunications policies use satellite phones to bypass local network restrictions, ensuring they can report freely and stay in touch with their home offices.
- Government and Military Use: Governments and military units around the world use satellite phones for field operations, especially in areas without secure or reliable communication infrastructure.

The usage of satellite phones involves dialing numbers in a similar manner to cell phones, though the dialing sequence can vary depending on the satellite network. Users must ensure a clear line of sight between the phone's antenna and the sky, as buildings, mountains, or heavy foliage can obstruct the signal. Despite these considerations, satellite phones are praised for their ability to provide critical communication links in the most challenging and remote environments on Earth.

In your emergency toolkit, a satellite phone is vital. It's not just another gadget. When local networks are down, this device connects you to the global satellite grid. This isn't luxury; it's a lifeline.

Selecting and Using Your Satellite Phone: A Quick Guide

1. Research and Choose Wisely: Start with thorough research on the best satellite phones available. Consider durability, battery life, coverage areas, and subscription fees.
2. Understand the Operations: Before an emergency strikes, familiarize yourself with the phone's functions. Know how to send and receive messages, make calls, and if available, use its GPS functionality.
3. Test Your Device: Conduct regular tests in different environments to ensure you can connect when it matters most. This also helps you get comfortable with the device's interface and features.
4. Keep It Charged and Ready: Always have your satellite phone charged and within reach. Consider investing in solar chargers or extra batteries for extended power options.
5. Secure and Waterproof Storage: Protect your satellite phone from the elements with a waterproof case or bag, especially if you're venturing into harsh environments.
6. Share Your Number: Make sure your family, friends, and emergency contacts have your satellite phone number. In crises, communication is key.
7. Regular Updates: Keep your phone's software updated and check for any service changes with your provider. This ensures your device remains reliable.
8. Emergency Contacts List: Store important numbers in the phone's memory, including local emergency services, your emergency contacts, and any relevant medical information.

Consider real scenarios where this technology has been crucial. Picture hikers in the Andes, isolated by a fierce blizzard. Traditional communications failed them. But one hiker carried a satellite phone. Despite the severe conditions, they reached out for help, sharing their location with rescuers. This decision, this device, was their salvation.

Satellites offer more than just communication. They are a beacon of hope, a tool for survival, and a must-have for anyone serious about preparedness. In a world of uncertainties, they provide one thing you can count on.Another poignant example is the use of satellite GPS in the aftermath of natural disasters. In the chaotic aftermath of a massive earthquake, when landmarks are obliterated and roads are impassable, satellite GPS can guide rescuers to those in need and help coordinate relief efforts. This technology was instrumental in the wake of the Haiti earthquake, where it was used not only for coordination and mapping but also for tracking and distributing aid effectively.

The moral of these stories? In the realm of preparedness, overlooking the potential of satellites is a mistake none of us can afford to make. It's about harnessing the power of technology, about looking to the stars, not in wonder, but as a practical step towards ensuring our safety and connectivity, no matter what the world throws our way. As we delve further into the age of digital technology, let's not forget the tools that can truly make a difference when disaster strikes. Equip yourself, educate your family, and embrace the power of satellites. In the world of emergency preparedness, staying connected means staying alive.

Satellite devices vary greatly, and the right choice hinges on understanding these variations. Durability is non-negotiable. Your device must withstand rough handling and extreme weather. The battery life is equally crucial; in an emergency, recharging could be impossible. And bandwidth matters more than you might think – it dictates the volume and speed of your communications.

But here's the thing: satellite communication isn't just about survival; it's about smart, informed survival. It's about knowing everything about your gear. Testing your equipment in different scenarios, familiarizing yourself with its quirks and features, and understanding its limitations can make all the difference. And remember, in an emergency, your satellite device is more than a tool; it's your link to the rest of the world.

Now, consider the environment. Extreme temperatures, from scorching heat to freezing cold, can impact device performance. Your gear should be able to handle this range. Waterproofing is another factor. Whether it's rain, snow, or accidental immersion, your device should come out functioning. And let's not overlook visibility. In the harsh glare of daylight or the dim of night, your device's display needs to be clear and legible.

Think about your location's peculiarities. Dense foliage, high walls, and deep valleys all interfere with satellite signals. If you're heading into such areas, choose devices known for strong signal capabilities. A portable satellite antenna might seem like an extra burden, but in the right circumstances, it can be the difference between isolation and connection.

Once you've secured the right equipment, maintaining a strong satellite connection becomes your next challenge. Start with power management. Battery life is finite, especially when you're off the grid. Extend it by reducing screen brightness, disabling unused features, and employing power-saving modes. Familiarize yourself with these adjustments long before you're in a critical situation.

A Satellite Drill

I urge every reader to consider conducting a "satellite drill" at least once a year. This involves simulating a scenario where traditional communication methods have failed, and you must rely solely on your satellite technology for communication. Ensure that every member of your family or group knows how to operate the satellite phone, send distress signals, and pinpoint their location using GPS coordinates.

Begin by selecting a day when you can afford to turn off all cell phones, internet connections, and traditional communication tools. Pretend that a major disaster has occurred, and your only link to the outside world is your satellite device. Practice making calls, sending text messages, and updating your location via satellite. Familiarize yourself with all the features of your device, including how to conserve battery life and what to do if the device malfunctions.

This drill will not only make you more comfortable and proficient with your satellite technology but also highlight any gaps in your preparedness plan. Remember, in a real emergency, you won't have the luxury of leisurely figuring out how to use your satellite tools.

The middle of a crisis is not the time for a crash course. So, take this advice to heart: Prepare, practice, and stay connected. It's a simple step, but it could very well be the one that saves your life or the lives of others when disaster strikes.

When you're picking out satellite communication gear, it's not a time for whims or impulses. This is about your safety, your lifeline in moments when conventional systems fail. Start by considering the essentials: what your specific needs are and the typical conditions of your environment. This isn't about owning the latest or most advertised gadgets; it's about what will serve you best in a crisis.

Choosing and maintaining satellite communication gear is a critical aspect of preparedness. It's not merely about having the right tools; it's about ensuring they'll work when you need them the most. Practical, thoughtful selection combined with rigorous, routine maintenance forms the backbone of effective emergency communication. In the end, your gear's reliability can significantly influence your survival and success in off-grid situations.

Battery Powered Devices

Cold climates pose additional challenges for battery-powered devices. Batteries drain faster in low temperatures. Counter this by keeping your devices insulated and close to your body when not in use. Body heat can prevent battery levels from plummeting.

Cold climates aren't the only issue for battery-powered devices; heat can be just as damaging. Batteries exposed to high temperatures can suffer from decreased capacity and longevity. To mitigate this, avoid leaving your devices in direct sunlight or in a hot car. Use insulated cases that can protect them from extreme temperatures, both hot and cold.

Moisture and water exposure pose significant risks to electronic devices, especially in outdoor or survival situations. Waterproof bags or cases are essential, but you can also use simple zip-lock bags in a pinch. For devices not in use, storing them in airtight containers with desiccants can absorb any moisture in the air, keeping the devices dry.

Dust and debris can clog ports, degrade interfaces, and even scratch screens, reducing visibility and functionality. Regular cleaning of your devices is crucial, using soft cloths and, if necessary, compressed air to blow out dust from harder-to-reach areas. Be particularly mindful of the charging ports, as debris here can prevent proper charging.

Signal interference can come from unexpected sources, including trees, buildings, and even the terrain itself. Understanding the direction from which your signal originates can help in positioning your device more effectively. In some cases, external antennas or signal boosters can enhance connectivity, especially if you're operating from a fixed location for an extended period.

Battery conservation is a continual concern. Besides environmental factors, the way you use your device can drain the battery. Limit the use of high-power functions like the screen, GPS, and data connections when not necessary. Turning off these features can extend your device's battery life significantly. Also, consider carrying multiple power sources, such as spare batteries or power banks, and remember that solar chargers can offer a sustainable charging solution in sunny conditions.

Technical glitches and software issues can also disrupt your communication tools. Regular updates can often resolve these problems, but in the field, resetting your device can be a quick fix for many minor issues. Familiarize yourself with the reset procedures for your devices, and back up important data to an external source whenever possible to prevent loss.

Privacy and security concerns with digital communication tools are ever-present. Using encrypted services can protect your information, but be aware of the vulnerabilities of each platform. Regularly update your passwords and use multi-factor authentication whenever possible to add an extra layer of security.

A strong signal is vital but often tricky in adverse conditions. Your positioning can significantly impact connectivity. Always seek a location with an unobstructed view of the sky. Even in seemingly open areas, small adjustments can dramatically improve signal strength.

Apps and gadgets

In the realm of survival, your smartphone isn't just a device for social media or casual browsing. It transforms into a crucial survival tool with the right applications. Think of it as a compact, digital Swiss Army knife. Let's dive into the essential apps that could save your life.

Navigation is your starting point. Forget traditional maps; digital mapping apps can pinpoint your exact location and guide you to safety. Look for apps that offer offline maps and real-time GPS tracking. It's not just about knowing where you are, but where you need to go, especially when traditional signposts are no longer an option.

Next up, first aid. A comprehensive first aid app is non-negotiable. In emergencies, medical help might not be immediately available. An app that provides step-by-step instructions for treating injuries can make the difference between life and death. Look for apps created or endorsed by reputable health organizations.

Emergency alerts are your information lifeline. Choose an app that provides real-time alerts for severe weather, natural disasters, and other emergencies. Staying informed means staying one step ahead of the game. Ensure the app is configurable to your specific location and needs.

But let's move beyond the screen. Your digital toolkit should extend to physical gadgets as well. Water purification devices have gone high-tech. Consider gadgets that use ultraviolet light to purify water, making it safe to drink in minutes. These devices are compact, easy to use, and invaluable when clean water is scarce.

Don't overlook the importance of staying informed. An emergency radio that receives AM/FM, shortwave, and NOAA weather alerts can be a lifeline in disaster scenarios. Modern versions come with hand cranks, solar panels, and USB charging ports, ensuring you stay connected even when the power's out.

Lastly, let's talk about multitools, but not just any multitools. The latest versions integrate digital features like USB chargers, emergency beacons, and GPS locators. It's the traditional survival tool, updated for the digital age.

Now, consider the story of a family stranded during a flash flood. Roads washed out, power lines down, they were isolated. But they weren't unprepared. With a navigation app, they found a safe route to higher ground. Their first aid app guided them in treating minor injuries. Alerts kept them informed of the flood's progression. Their water purifier ensured access to clean drinking water, the emergency radio kept them updated, and their digital multitool charged their devices, ensuring continuous communication.

This is not just a hypothetical scenario. It's a real possibility, and with the right apps and gadgets, survivability increases exponentially. Your digital devices are not just conveniences; in the hands of a prepared individual, they are lifelines. So arm yourself with knowledge, equip your devices wisely, and embrace the fusion of traditional survival skills with modern technology. Remember, in the world of emergency preparedness, being informed and equipped is not just an advantage, it's a necessity.

In today's world, integrating technology into your survival plan isn't just smart; it's essential. The right apps and gadgets can enhance your situational awareness and sharpen your decision-making in critical times. Think of these tools as extensions of your survival instincts, digital eyes and ears that help you perceive and react to your environment with greater precision.

Let's start with integrating apps into your plan. Your smartphone, often dismissed as a daily convenience, is a powerhouse of survival resources. Install and familiarize yourself with key apps before disaster strikes. Navigation apps can guide you out of danger zones, while first aid apps provide crucial medical information when help isn't on the way.

Weather and emergency alert apps keep you informed about your surroundings, giving you precious time to react. But remember, knowledge without practice is like a lamp without oil. Regularly review these apps, practice scenarios, and ensure your family knows how to use them too.

Now, onto the gadgets — your physical tools. A solar-powered charger, for example, is invaluable when power sources are scarce. But it's not just about having the gadgets; it's about knowing where they are and how to use them efficiently. Regular checks are crucial; ensure everything is in working order and that you're comfortable using each device. Incorporate them into your drills. Familiarity breeds efficiency.

Keeping your devices operational is just as crucial as having them. Power management becomes paramount when you're off-grid. Extend your device's battery life by adjusting settings: dim your screen, switch to power-saving modes, and close unnecessary apps. But don't stop there. Invest in portable power sources like solar chargers or power banks. And remember, these devices need to be kept safe from the elements too. Waterproof cases and protective pouches can shield your electronics from water, dust, and impact.

But what does this look like in your real life? Consider the case of a family caught in a hurricane's path. With power out, their smartphones, equipped with emergency apps, became their primary source of updates. Alerts received gave them enough warning to evacuate safely before the worst hit. Throughout their ordeal, their solar chargers kept their phones operational, ensuring continuous access to weather updates and enabling them to communicate with emergency services.

The family had not only chosen the right gadgets but had also taken the time to integrate them into their emergency plan. They knew how to manage their devices' power needs and had taken steps to protect them from the storm's impacts.

Understand this: when chaos unfolds, information becomes as crucial as water or shelter. But it's not just about sending and receiving messages. It's about ensuring that those messages remain confidential, that they reach only those for whom they are intended. The consequences of compromised communications range from the inconvenient to the dangerous. In crises, information leaked to the wrong people can lead to looting, ambushes, or worse.

Now, let's talk encryption. It sounds technical, and it is, but here's the bottom line: encryption transforms your messages into a code. The only way to decipher this code is with a unique key. Imagine sending a locked box through the mail. Without the key, the contents remain a mystery. This is what encryption does for your communications.

In today's digital age, various tools offer encryption. Messaging apps with end-to-end encryption provide a solid start. These ensure that only you and your recipient can read what's sent. Nothing gets revealed to third parties, not even to the companies that host these services.

But how do you incorporate these tools into your preparedness plans?

- Identify Robust Encryption Tools: Start by pinpointing apps and services known for their strong encryption.
- Do Your Homework: Dive into research. Read reviews and explore the features of each tool to ensure they meet your security needs.
- Test Thoroughly: Once you've chosen your tools, test them extensively. Send test messages, make calls, and use any other available features to ensure they work as expected.
- Learn the Ropes: Become intimately familiar with how each app operates. Know how to navigate its settings and use its features effectively.
- Train Your Circle: Make sure your family and close allies are also well-acquainted with these tools. They should be as comfortable using them as you are.
- Practice Regularly: Incorporate the use of encrypted messaging into your emergency drills. Send messages back and forth to ensure everyone can communicate smoothly under pressure.

Prepare for Quick Use: Remember, in an emergency, there's no time to waste. Ensure you and your network can use these apps swiftly and without hesitation.

Consider the case of a community facing a sudden evacuation due to a wildfire. Panic spread, misinformation was rampant. But one group, having set up encrypted communication channels in advance, was able to share verified information, coordinate safe routes, and assist each other without broadcasting their plans to everyone. Their preparedness and understanding of secure communication tools allowed them to navigate a fraught situation with added security and clarity.

This isn't just a hypothetical. In every emergency, from natural disasters to man-made crises, the security of your communication can significantly impact your safety and survival. Whether it's coordinating a meetup point or sharing the status of your supplies, you want to ensure that this information stays secure and doesn't fall into the wrong hands.

Let's be clear: adopting secure communication practices is not about paranoia; it's about preparedness. In an age where information can be as valuable as physical assets, protecting your communications is critical. Start simple. Educate yourself and your network on encryption. Practice its use. Make secure communications a staple of your emergency plans.

Just as you wouldn't leave your physical doors unlocked, don't leave your digital doors open. In the landscape of survival, the strength of your preparations extends beyond the physical to the digital realm. Your vigilance here, your attention to the confidentiality and integrity of your communications, can make all the difference when the unexpected hits.

It's essential that you use secure channels for your communication. This means platforms that offer end-to-end encryption, ensuring that only you and your intended recipient can decipher the messages. But it's not just about choosing the right platform; it's also about using it correctly.

Always verify the identities of your contacts. Be cautious about sharing sensitive information, even on encrypted platforms. And remember, the strongest encryption can be undone by a weak password. Use complex and unique passwords and remember to change them regularly.

Now, let's tackle common security pitfalls. One major mistake is complacency. Just because a conversation is encrypted doesn't mean you're invincible. Be wary of phishing attempts and malware. These can compromise your device and, by extension, your communications. Another pitfall is overconfidence in public Wi-Fi. These networks are hotbeds for eavesdropping. If you must use public Wi-Fi, employ a reputable virtual private network (VPN) to encrypt your internet traffic.

Understanding the legal landscape surrounding encrypted communications is crucial. It is important to know that laws may vary significantly depending on the jurisdiction. In general, using encryption is legal, but there are contexts and jurisdictions where it can raise flags or require compliance with specific regulations. Educate yourself about the laws in your area and any area you might travel to. Ignorance of the law is not a defense, especially when your safety and privacy are on the line.

Ethical use of encrypted communications is equally important. Just as you value your privacy, respect the privacy of others. Use encryption responsibly and in a way that does not harm others. Avoid spreading misinformation or using encrypted channels for illicit activities. Remember, the goal is to enhance your safety and the safety of your community, not to undermine it.

A checklist for secure communication:

1. Choose Secure Communication Platforms: Opt for messaging and email services that offer end-to-end encryption.
2. Verify Contacts: Always confirm the identity of your contacts on encrypted platforms to avoid communicating sensitive information to the wrong person.
3. Strengthen Your Passwords: Use complex, unique passwords for your accounts and change them regularly to enhance security.
4. Beware of Public Wi-Fi: Avoid transmitting sensitive information over public Wi-Fi networks. Use a VPN to encrypt your internet traffic when you must use public networks.
5. Stay Informed on Legalities: Educate yourself about the legal frameworks regarding encrypted communications in your jurisdiction and any areas you plan to visit.
6. Practice Ethical Communication: Use encryption responsibly, respecting the privacy of others and avoiding illicit activities.
7. Regularly Update Your Knowledge: Stay informed about the latest security threats and updates for your chosen communication platforms.

Implement Anti-Phishing Measures: Be vigilant against phishing attempts by not clicking on suspicious links and verifying the authenticity of messages received.

Consider the case of a group of activists who used encrypted messaging to coordinate relief efforts after a natural disaster. By employing secure communication practices, they were able to share sensitive information about the locations and needs of affected individuals without exposing them to additional risks. This approach not only protected the activists from surveillance but also ensured the safety and privacy of the people they were helping.

Vigilance, awareness, and a commitment to privacy and security are the cornerstones of secure communication. Whether you're navigating the aftermath of a disaster, coordinating with your preparedness community, or simply going about your daily life, these practices can protect you from a myriad of threats.

Portable Solar Technology

In today's survival toolkit, portable solar technology is a must-have. It's efficient, durable, and crucial for keeping your devices powered up when you're off-grid. Let's break down what to look for and how to use these tools effectively.

First up, choosing the right gear. Your solar panels should be foldable, lightweight, and efficient. Efficiency matters because you want to capture as much sunlight as possible, even on cloudy days. Durability is also key. Your gear should withstand rough weather and be water-resistant.

For portable chargers and power banks, focus on features useful in survival scenarios. Some come with built-in LED lights, emergency SOS signals, and multiple ports for charging different devices at once. Capacity is critical. You'll want a power bank that can charge your phone or radio several times over.

To use these devices, you need sunlight and patience. Position your panels towards the sun for best results. Remember, charging with solar energy is slow. Keeping the panels clean helps them absorb more sunlight.

Maintenance is simple but necessary. Protect your panels from extreme heat to avoid damage. Regularly check cables and connectors for wear. Replace them if needed to ensure a steady charge.

Imagine being stranded with no power. With a portable solar charger, you can keep your critical devices like satellite phones charged. This isn't just convenient; it can be lifesaving, allowing you to call for help even from remote locations.

In summary, portable solar tech is not an optional luxury. It's a vital part of your survival gear. Choose efficient, durable equipment. Use it wisely. Maintain it regularly. This way, you'll ensure you have power when and where you need it most. Solar power isn't just about being eco-friendly; it's about being prepared.

A mesh network is like a grid of devices that talk directly to each other instead of relying on a central hub like a Wi-Fi router or cell tower. Picture a web where every phone or laptop is a node connected by virtual threads. This setup is perfect for situations where traditional communication lines are down because it creates a standalone network that can operate entirely off the grid.

The reason people lean towards mesh networks, especially in survival scenarios, is their resilience. If one device drops off, the network self-heals by rerouting through other devices, ensuring messages still get through. They're used for everything from sending emergency texts within a group to sharing real-time info across a disaster-hit community. The beauty of a mesh network lies in its simplicity and its strength—the more devices join, the stronger and further the network reaches.

Setting up a mesh network is straightforward. You'll need devices capable of connecting to each other—these can be smartphones, laptops, or specific mesh-enabled devices. The key is using software or apps designed for mesh networking. These apps allow your devices to send and receive messages through the network of connected devices, extending the range beyond a single device's capability.

Here's how to get started:

- Choose Your Devices: Any device with Wi-Fi or Bluetooth can be part of your mesh network. Ensure they're charged and ready.
- Select a Mesh Networking App: Several apps are available for creating mesh networks. Download the same one on all devices to ensure compatibility.
- Set Up Your Network: Follow the app's instructions to connect your devices. This usually involves turning on Bluetooth or Wi-Fi and allowing the app to connect to other nearby devices.
- Test Your Network: Send messages between devices to check the network's range and reliability. See how far apart devices can be while still maintaining a connection.
- Maintain Your Network: Keep your devices charged and within range of each other. Remember, the more devices, the stronger and wider your network.

Consider a case where a group of hikers uses a mesh network to stay connected in an area with no cell service. Each hiker's smartphone is part of the network, allowing them to send updates and locations to each other, even when spread out. This setup provides a lifeline, ensuring no one is left out of reach.

Mesh networks aren't just for tech experts; they're a practical tool for anyone needing reliable communication off the grid. With a bit of preparation, you can establish your network, keeping your group connected when it matters most.

This guide has taken you through the crucial skills and tools needed to maintain communication in the absence of traditional networks, marrying hands-on advice with cutting-edge solutions.

We've explored the transformative power of satellite communications in crisis situations, unveiled essential apps and gadgets that could be lifesavers, and illustrated how to weave your own mesh network using everyday technology. The goal was to arm you with the strategies and insights necessary to keep you ahead of the curve, regardless of the challenges you face.

From understanding the ins and outs of satellite gear tailored to your needs, to harnessing solar power to keep your devices charged, we aimed to keep it practical, direct, and useful. The focus on securing your communications wasn't just about tech-savviness; it was about ensuring privacy and safety in times when it matters most. With these strategies and tips, you're now better equipped to handle the challenges of off-grid communication, making sure you, your loved ones, and your community can remain connected and secure, whatever the world throws your way.

Book 9: Beyond Words: Non-Verbal Communication and Community Network.

In the world of emergency preparedness, being able to send a message that only your group understands could be what keeps you safe. This is where cryptography comes into play. It's more than old spy movies; it's a real skill you can use to keep your communications private, especially when you can't rely on regular channels.

Cryptography is the art of coding and decoding messages, turning your information into a puzzle only those you trust can solve. Imagine you need to send a message about your location or a planned move, but you don't want just anyone to understand it if they intercept it. By using a simple cipher, you can encode this message, making it appear as a string of unrelated letters or numbers to anyone outside your group.

Here's how you get started with the basics: First, pick a cipher. The Caesar cipher, a method where each letter in your message is shifted a certain number of steps down the alphabet, is a straightforward option. If your agreed-upon shift is three, then every 'A' becomes 'D', 'B' becomes 'E', and so on through the alphabet. If we reach the end, we circle back to the beginning (so, X becomes A, Y becomes B, and Z becomes C).

Let's encode the message: "MEET AT DAWN."

Following our cipher rule, shifting each letter three places:

M becomes P

E becomes H

E becomes H

T becomes W

A becomes D

T becomes W

D becomes G

A becomes D

W becomes Z

N becomes Q

So, "MEET AT DAWN" encoded with a Caesar cipher with a shift of three becomes: "PHHW DW GDZQ."

To decode the message, you simply shift in the opposite direction, moving three places up the alphabet, turning "PHHW DW GDZQ" back into "MEET AT DAWN."

Remember, the key to using this code effectively within your group is making sure everyone knows the agreed-upon shift number. Practice encoding and decoding messages to become fluent in using your cipher.

For practical applications, consider using coded messages for critical but sensitive information, like sharing safe routes, locations of supplies, or meetup points. The beauty of cryptography is its versatility; you can use it in any scenario, urban or wilderness, ensuring your group stays informed without compromising security.

Now, let's talk tools and techniques. You don't need anything fancy; a simple piece of paper and pen can suffice for writing down and deciphering codes. For a more modern approach, various apps and software can generate and decode ciphers, but remember, the goal is to stay operational even when tech isn't available.

Cryptography is a powerful tool in your emergency preparedness arsenal. It allows for secure communication in situations where discretion is paramount. By understanding the basics of creating and deciphering codes, and by practicing regularly, you ensure that when the need arises, you're ready to use cryptography to protect your communications, keep your group safe, and stay one step ahead in any situation.

Sign language, emergency signals and symbols

Imagine you're in a scenario where silence is paramount—perhaps hiding from a threat or moving undetected through an area. Here, basic sign language can be a lifesaver. Essential signs such as "Danger," "I need help," "Food," and "Water" can be communicated silently within your group. Learning these signs isn't just about expanding your skill set; it's about ensuring mutual understanding and support without uttering a single word. Practice these signs with your group, making sure each gesture is clear and universally understood among your members.

The use of flares, lights, and flags as distress signals is a time-tested method of communicating your need for help across distances. A bright flare shot into the sky can be seen for miles, instantly alerting rescuers to your location. Similarly, using flashlights or even mirrors to send an SOS signal—three short flashes, three long, three short—can catch the attention of search teams or passersby. And don't underestimate the power of brightly colored flags or cloths; placed in open areas, they can mark your position for aerial searchers.

In both urban and wilderness settings, marking safe paths or hazardous zones with universally recognized symbols can guide others to safety or steer them clear of danger. A simple "X" can signal danger, while an arrow can point to safe passage. These symbols can be created with spray paint, etched into surfaces, or even arranged with stones or branches. The key is consistency and visibility. Agree on the meaning of each symbol with your group and ensure they're prominently placed where they can be easily seen.

Consider the story of a group lost in a dense forest. With night falling and one member injured, verbal communication was risky due to the presence of wildlife. Using basic sign language, they decided to stay put. They marked their location with a large "SOS" symbol on the ground, made visible from the air with whatever was on hand—clothes, branches, and a flashlight set to flash SOS in Morse code. A flare was kept ready for when they heard rescue teams nearby. Their silent signals paid off. They were found by a search party that spotted the SOS from a helicopter, with the flare confirming their exact location.

In emergencies, your ability to communicate without words can make all the difference. By mastering sign language basics, utilizing emergency signals effectively, and marking your environment with clear symbols, you ensure that you can call for help, guide others to safety, or navigate danger without making a sound. This silent skill set is invaluable, empowering you and your community to face challenges with confidence and discretion.

Joining Forces and Creating a Survival Network.

Building a survival network is about more than just self-preservation; it's about leveraging the collective strength of your community to weather any storm. The foundation of this network isn't just about who you know; it's about making strategic connections and preparing together for whatever may come. Let's break down how to effectively establish this network and communicate within it, without relying solely on words.

Start small. Your first step isn't a grand assembly; it's about fostering trust and cooperation among neighbors. This might mean helping out with small tasks or sharing resources. It's these acts of goodwill that lay the groundwork for stronger ties. Once you've built up a level of trust, introduce the idea of a preparedness network. Use whatever platforms you have at your disposal—social media, community boards, or simple face-to-face conversations. The aim is to get everyone on the same page about the value of a collective survival strategy.

With a group formed, it's time to identify everyone's strengths. Who's the medical expert? Who knows how to fix just about anything? Assign roles based on these skills. This ensures the group is well-rounded and prepared for various scenarios. It's also critical to establish clear leadership for coordination, but maintain an environment where every voice can be heard and respected.

Silence can be a strategy. Develop a system of non-verbal signals or symbols for communication. This could be something as simple as a colored ribbon system: green for "all clear," red for "urgent help needed." These signals should be straightforward and universally understood within your group.

Take, for instance, a community facing a massive power outage after a storm. Before the storm hit, they had agreed on a flag system. Green flags meant "safe," yellow was for "need help but not urgent," and red was for "emergency." This simple yet effective method allowed them to quickly assess needs and mobilize support without unnecessary exposure or risk.

In crafting your survival network, remember that the goal is unity and resilience. By establishing solid connections, recognizing each member's unique contributions, and setting up a reliable method of silent communication, you're not just preparing to survive; you're ensuring the collective welfare of your community. It's this proactive, prepared mindset that transforms individuals into a formidable, cohesive unit ready to face any challenge head-on.

The Art of Silent Coordination

Mastering the art of silent coordination is a critical skill set for anyone serious about survival and self-reliance. It's not just about stealth; it's about effective communication and movement without making a sound. Whether you're navigating through potentially hostile territory or simply need to maintain silence to avoid drawing attention, these techniques can be lifesaving.

Developing a set of hand signals is your first step. These need to be simple, clear, and understood by every member of your group. Start with the basics: stop, go, look, danger. From there, tailor your signals to fit common situations you might encounter. For instance, signals for "I see water" or "Need help" can be invaluable. The key is consistency and practice. Everyone in your group should know these signals by heart.

Moving as a unit silently requires more than just following each other. It's about maintaining formation, distance, and being aware of each other's positions without needing to look back. Use a leader to guide the group's direction, with each member keeping the person in front of them in their peripheral vision. Practice moving in different terrains and conditions to understand how your spacing and pace might need to adjust.

The only way to ensure these techniques become second nature is through regular practice. Set up drills that simulate real-world scenarios. Start with simple exercises, like moving from point A to B silently in a controlled environment. Then, increase complexity by adding objectives, such as retrieving items or navigating around obstacles without breaking silence. Include scenarios that require quick decision-making through non-verbal communication, such as changing direction due to a sudden "threat."

Silent coordination is not just a tactic; it's a vital component of emergency preparedness and survival strategy. By developing and practicing hand signals and movement techniques, your group can enhance its ability to operate stealthily and effectively, whatever the circumstances. Remember, the success of silent coordination hinges on everyone's commitment to learning, practicing, and perfecting these non-verbal communication skills.

Reading the Signs: Environmental Cues

In the world of survival and emergency preparedness, being able to read the environment around you is an invaluable skill. Nature offers a myriad of signs that, if interpreted correctly, can guide your path, predict the weather, and warn you of impending dangers. Let's delve into how you can harness these environmental cues to your advantage.

The art of natural navigation involves using the stars, plants, and animals to find your way. For centuries, travelers have looked to the North Star in the Northern Hemisphere to find north, a fundamental skill for orientation during clear nights. During the day, observe the sun's path; it rises in the east and sets in the west, but its exact trajectory can tell you the time of year. Plants and animals also offer clues; for instance, moss often grows more heavily on the northern side of trees in the Northern Hemisphere, while birds tend to fly towards water sources at dawn and dusk.

Predicting the weather isn't just about comfort; it's about survival. Cloud formations, for instance, can tell you a lot about what the weather has in store. Cumulus clouds, fluffy and spread out, generally indicate fair weather, but when they start to pile up and darken at the base, it's a sign that storms may be brewing. Wind patterns are also telling; a sudden shift in wind direction can indicate an approaching weather front. Learning to read these signs can give you the upper hand, allowing you to seek shelter before conditions worsen.

Nature has its way of warning us about potential disasters. Before earthquakes, for instance, animals often act erratically, due to their sensitivity to the Earth's vibrations. A sudden drop in water levels in rivers or coastal areas can be a precursor to tsunamis. And, the smell of sulfur in volcanic regions can indicate that an eruption might be imminent. Being attuned to these signals can provide crucial minutes or even seconds to react.

Consider the story of a hiker who, while traversing a remote trail, noticed the clouds overhead shifting from scattered cumulus to towering cumulonimbus formations. Recalling the basics of weather prediction, he recognized this as a sign of an impending storm. Observing the wind's sudden change in direction, he quickly sought shelter under a rocky overhang, just as the skies opened up. His ability to read the signs nature provided not only kept him dry but likely saved him from being caught in a dangerous situation, with lightning striking nearby trees.

Mastering the skill of reading environmental cues is not just about enhancing your outdoor experiences; it's a fundamental aspect of staying safe and making informed decisions in the wild. By attuning yourself to nature's language—whether navigating by the stars, predicting the weather, or heeding hazard signals—you empower yourself with knowledge that no gadget can replace. This awareness, combined with practice and application, ensures that you can confidently navigate the challenges that nature may present.

Here's how you can begin to integrate these skills into your preparedness repertoire.

Steps for Natural Navigation

- Learn the Stars: Familiarize yourself with key constellations and how they change with seasons. Practice locating the North Star (Polaris) or the Southern Cross as reliable guides.
- Observe Wildlife: Pay attention to the behavior of birds and insects. Many species have patterns that can indicate the direction of water or changes in the weather.
- Study Tree Growth: Notice the side of trees where moss grows more abundantly, which can indicate north in the Northern Hemisphere, helping you to orient yourself.
- Techniques for Weather Prediction
- Cloud Watching: Spend time observing different cloud types and their movements. Learn the differences between stratus, cumulus, and cirrus clouds and what each type can tell you about impending weather.
- Wind Direction: Create a simple wind vane or use a wet finger to determine wind direction. A sudden change can often signal an approaching storm front.

Pressure Drops: If you have a barometer, monitor it closely. A rapid drop in pressure is a clear sign of worsening weather conditions, prompting you to seek shelter.

Identifying Hazard Signals:

- Animal Behavior: Keep a journal of animal behavior observations during your outdoor adventures. Over time, you'll start to notice patterns that could indicate natural events like earthquakes or storms.
- Water Level Changes: Always note the water level of rivers or coastal areas when you set up camp. A noticeable drop could indicate a tsunami or flash flood, giving you time to move to higher ground.
- Sulfur Smell: In volcanic regions, be hyper-aware of the smell of sulfur or a sudden increase in ground temperature, both indicators of volcanic activity.
- Putting It All Together: A Practical Exercise

Choose a day to go out into a natural setting, preferably one with a variety of landscapes. Leave all digital devices behind, and bring a notebook, a compass, and if possible, a basic weather observation kit.

- Morning: Start by observing the sky and noting the cloud types and wind direction. Predict the weather for the next few hours.
- Midday: Focus on navigation. Use the position of the sun to determine the right direction. Look for natural signs to confirm your findings, like moss on trees or the behavior of wildlife.
- Evening: As night falls, identify as many constellations as you can and use them to find your bearings. Note any changes in animal activity that could indicate changes in weather or other environmental factors.
- Reflect on this experience. What matched your predictions and observations? What didn't?

This practice will sharpen your skills and increase your confidence in using environmental cues for survival.

By actively engaging with and learning from your environment, you become better equipped to anticipate and react to changes, ensuring you remain steps ahead in any survival situation. These skills, developed over time, can transform the natural world into a guide, ally, and warning system, all without saying a word.

Emotional Intelligence: The Silent Communicator

Emotional intelligence, especially in survival situations, becomes a silent but powerful communicator. Understanding and interpreting body language, maintaining morale, and resolving conflicts non-verbally are skills that can significantly impact your group's dynamics and overall success.

In high-stress environments, the ability to read someone's body language can give you insights into their true feelings or intentions, often before they even say a word. A person clenching their fists or avoiding eye contact might be experiencing stress or hiding something. Conversely, relaxed shoulders and open posture can indicate comfort and trust. Paying attention to these subtle cues allows you to address concerns before they escalate, ensuring everyone's on the same page.

Morale is the backbone of any effective team, especially in survival situations. Non-verbal cues like a pat on the back, a thumbs-up, or even a smile can significantly boost someone's spirits. Establishing a culture of positive reinforcement through gestures and expressions fosters a sense of unity and support. Celebrate small victories together with shared gestures, like a group high-five, to reinforce team cohesion.

Conflicts are inevitable, but how you handle them can make all the difference. Non-verbal cues can play a crucial role in de-escalating tensions. Adopting a calm demeanor, using open body language, and maintaining non-threatening eye contact can help soothe tempers without a word. Sometimes, simply sitting side by side with someone, rather than facing them directly, can make a conversation feel less confrontational, facilitating easier resolution.

- Practice Observing: Spend time each day observing people's body language. The more you practice, the better you'll get at reading non-verbal cues accurately.
- Non-Verbal Communication Drills: Incorporate non-verbal communication exercises into your group's training sessions. For example, try completing a task silently, using only gestures to coordinate.
- Role-Playing Conflict Scenarios: Simulate conflict situations and practice resolving them non-verbally. This helps prepare everyone for handling real-life tensions effectively.

Consider a situation where a group member starts isolating themselves, showing signs of stress through body language but not speaking up. Recognizing these signs early allows the group to intervene, offering support through simple, comforting gestures or sitting down beside them in a non-confrontational manner to encourage sharing. Such proactive measures can prevent stress from turning into a full-blown crisis, maintaining group harmony and morale.

Emotional intelligence is a critical, yet often overlooked, component of survival strategy. It goes beyond mere survival tactics, touching on the heart of what it means to work as a cohesive unit. By honing your ability to communicate silently, read between the lines, and maintain a positive group dynamic, you equip your team with the resilience to face not just the physical challenges of survival, but the emotional and psychological ones as well.

Technology in Silence: Digital and Non-Verbal Tools

In the world of emergency preparedness and off-grid communication, technology stands as a reliable ally, especially when words can't convey what's needed. For folks like us, practical and hands-on, who prioritize self-reliance in emergencies, understanding and utilizing digital and non-verbal tools can be a game-changer for staying safe and connected when verbal communication isn't an option.

The concept of non-verbal communication through technology has been around for a while, inspired by the emergency distress signals used in maritime and aviation industries. As our smartphones became more advanced, so did these apps, evolving to cater specifically to emergency preparedness and off-grid communication.

The idea of wearable tech isn't new—it traces back to things like pocket watches. But today, wearable tech has stepped up its game with cutting-edge features like GPS, Bluetooth, and cellular connectivity. Now, we've got smartwatches, bracelets, and pendants that discreetly keep us connected, even in the toughest situations.

Light and laser signals are like our silent beacons, guiding others to us when words fail. It's about using beams of light to send coded messages across long distances. This technique has been used for ages in various fields, from maritime navigation to military operations and search and rescue missions.

The history of light and laser signals goes back to Morse code, a system of communication using dots and dashes. But with today's technology, we've taken it up a notch with laser pointers and LED flashlights, making our signals clearer and more visible than ever.

In the end, technology in silence is all about having the right tools and know-how to communicate effectively when words aren't enough. Whether it's through non-verbal apps, wearable tech, or light and laser signals, these innovations empower us to stay connected and call for help when we need it most. By embracing these digital and non-verbal tools, we're not just preparing for emergencies—we're ensuring our safety and resilience in the face of adversity, even when silence surrounds us.

Traditions and Rituals: Unspoken Bonds

In many cultures, specific symbols and gestures hold profound meaning and can convey a wide range of messages without uttering a single word. For example, in some communities, a hand placed over the heart signifies sincerity or respect, while in others, it may indicate a need for assistance or distress. Understanding these nuances is critical for effective communication, particularly in diverse or multicultural settings.

Take, for instance, the Navajo Nation's use of the Navajo Sign Language (NSL), a complex system of hand signals and gestures that evolved as a means of communication among the deaf and hearing members of the community. By recognizing and respecting such cultural symbols, we not only enhance our ability to communicate across language barriers but also demonstrate respect for the traditions and customs of others.

Rituals have a profound impact on group cohesion and can serve as powerful mechanisms for building trust and solidarity within a community. Consider the tradition of "fire circles" among survivalist groups, where members gather around a campfire to share stories, skills, and experiences. These gatherings not only provide practical opportunities for learning and preparedness but also foster a sense of belonging and mutual support.

Furthermore, rituals can be tailored to specific contexts or objectives, such as emergency response training exercises conducted by volunteer organizations like the Community Emergency Response Team (CERT). These exercises simulate real-world scenarios and allow participants to practice essential skills while reinforcing the importance of teamwork and collaboration.

Heritage skills encompass a wide range of traditional knowledge and practices that have been passed down through generations. In the context of emergency preparedness, these skills include everything from navigation techniques and tracking to wilderness survival and non-verbal communication methods.

For example, the art of tracking—a skill passed down through indigenous cultures for centuries—involves interpreting subtle signs and indicators left by animals or humans. By teaching younger generations how to read tracks, recognize animal behavior patterns, and interpret natural signs, we ensure that these essential skills are preserved and passed on.

Cultural Symbols and Gestures: Learn the meanings behind specific symbols and gestures in diverse cultures to enhance cross-cultural communication and understanding. For instance, a hand over the heart may convey sincerity or a need for assistance in different cultural contexts.

Ways to Preserve Heritage and Strengthening Community Bonds:

1. Navajo Sign Language (NSL): Explore the complexity of NSL, a communication system within the Navajo Nation, to grasp the significance of non-verbal communication methods

within specific cultural groups.
2. Community Rituals for Cohesion: Participate in community rituals like "fire circles" to foster trust and solidarity while sharing practical skills and experiences within survivalist groups.
3. Emergency Response Training Exercises: Engage in training exercises like those conducted by CERT to simulate real-world scenarios, emphasizing teamwork and collaboration for effective emergency response.
4. Heritage Skills Preservation: Pass down heritage skills such as tracking, navigation, and wilderness survival to younger generations to ensure the preservation of traditional knowledge and practices.
5. Basic Signaling Techniques: Teach children and grandchildren basic signaling methods like whistle codes or flag semaphore to instill self-reliance and resourcefulness while strengthening family bonds and cultural heritage.

Traditions and rituals play a vital role in strengthening community resilience and preparedness. By understanding and respecting cultural symbols, establishing meaningful group rituals, and passing down heritage skills to future generations, we can enhance our ability to communicate effectively and navigate uncertain times with confidence and resilience.

Stealth and Surveillance: Watching Without Being Seen

Let's talk about something crucial: staying under the radar when things get hairy. In the world of emergency prep, knowing how to move quietly and observe without being spotted can be a game-changer. So, let's get down to brass tacks and cover the basics of stealth and surveillance—stuff that could keep you safe when the chips are down.

Listen, blending into your surroundings isn't rocket science, but it takes a bit of know-how. First off, dress the part—wear colors that match where you're at. Earthy tones like brown and green work wonders in natural settings. And don't be afraid to get a little dirty—mud, leaves, and branches can help break up your outline and keep you hidden.

But it's not just about your clothes; it's about how you move, too. Slow down, be quiet, and avoid making any sudden moves that might give you away. Use what's around you—trees, bushes, rocks—to keep out of sight. Get good at this, and you'll be like a ghost in the woods, slipping by unnoticed.

You need to keep your eyes peeled without making it obvious. The key here is to notice everything without staring like a deer caught in headlights. Use your peripheral vision to scan your surroundings without giving yourself away. Keep your head on a swivel, but keep it low—no need to make eye contact with everyone you see.

And when you are watching for trouble, do it on the sly. Don't be obvious about it—casually glance around, use your instincts, and trust your gut. You'll pick up on things that others might miss, giving you the edge when it counts.

Now, let's talk about staying off the grid when someone's trying to keep tabs on you. First off, watch your digital footprint—limit your use of phones and gadgets that could give away your location. And when you are out and about, mix things up. Don't take the same route every time, and don't fall into a predictable routine.

Keep an eye out for anyone who might be tailing you. Look for signs—a car that's been following you too long, someone hanging around a little too close. And if you think you're being tracked, don't panic. Change up your plans, throw in a few curveballs, and shake off anyone who might be on your trail.

steps for mastering stealth and surveillance:

1. Blend In: Wear earthy tones like brown and green to match your surroundings. Use natural materials like mud, leaves, and branches to break up your outline and remain hidden.
2. Move Silently: Slow down and avoid sudden movements that might attract attention. Utilize natural cover such as trees, bushes, and rocks to stay out of sight.
3. Observe Discreetly: Use your peripheral vision to scan your surroundings without drawing attention. Keep your head low and avoid making direct eye contact with others.
4. Trust Your Instincts: When watching for trouble, remain subtle and rely on your intuition. Casually glance around, and use your instincts to pick up on potential threats.
5. Minimize Digital Footprint: Limit the use of phones and gadgets that could reveal your location. Be mindful of your online presence and avoid sharing unnecessary information.
6. Stay Unpredictable: Vary your routes and routines to avoid being tracked. Keep an eye out for anyone who may be following you and be prepared to change plans if needed.
7. Use Decoys and Distractions: Employ tactics like leaving false trails or dropping breadcrumbs to throw off pursuers. Keep adversaries guessing and maintain the upper hand.

Sometimes your best defense is a good strong offense. Throw them off your scent with decoys and distractions—leave false trails, drop a few breadcrumbs, and keep them guessing. It might sound like something out of a spy movie, but in the real world, a little misdirection can go a long way.

In a nutshell, mastering stealth and surveillance is all about staying one step ahead. Whether you're slipping through the woods or navigating a crowded city street, knowing how to move unseen and stay off the radar could mean the difference between staying safe and getting caught. So keep your wits about you, trust your instincts, and remember: sometimes, the best way to be seen is to stay hidden.

The Future of Non-Verbal Communication: Innovations on the Horizon

Alright, let's take a glimpse into what lies ahead for non-verbal communication in the world of emergency preparedness and off-grid living. For practical folks like us, staying ahead of the game means keeping an eye on emerging technologies, exploring the potential of artificial intelligence (AI), and understanding the importance of non-verbal communication in our increasingly connected global community.

Picture this: a future where communication flows effortlessly, even when words aren't an option. That's the promise of emerging technologies in non-verbal communication. We're talking about wearable devices that can interpret our gestures and translate them into actions, or smart sensors that detect distress signals and summon help without a spoken command.

Imagine a scenario where a lost hiker signals for assistance with a simple hand gesture, instantly picked up by drones equipped with advanced AI. Or a survivor trapped in a disaster zone, using a wearable device to communicate their location and condition to rescue teams in real-time. These innovations have the potential to redefine how we communicate in emergencies.

Now, let's talk about the role of artificial intelligence in shaping the future of non-verbal communication. AI has the power to transform how we interact with technology, enabling devices to understand and respond to our gestures, expressions, and body language in ways previously unimaginable.

Imagine a smartphone that detects signs of distress in your voice or facial expressions and automatically alerts your emergency contacts. Or a home system that anticipates your needs and adjusts the environment accordingly, all without you uttering a word. These advancements in AI-driven non-verbal communication could significantly enhance our safety and convenience.

In our ever-connected world, non-verbal communication serves as a universal language, bridging cultural and linguistic divides. Whether through gestures, facial expressions, or body language, these cues convey meaning and intention across borders and languages.

Think about it—a simple raised hand signaling distress transcends cultural differences, serving as a beacon for help. As we continue to build a global community, understanding and respecting non-verbal communication become paramount.

When we consider the intricacies of non-verbal communication and community network-building, it's evident that the path to preparedness is multifaceted and ever-evolving. From mastering the art of cryptography to harnessing the power of emerging technologies, we've explored a wide array of tools and techniques for communicating without words.

Throughout this book, we've learned not only how to decipher codes and signals but also how to forge bonds within our communities that can withstand the test of time. By understanding the importance of cultural symbols, rituals, and heritage skills, we've seen how tradition can serve as a powerful force for unity and resilience.

Looking ahead, the future of non-verbal communication holds boundless possibilities. With emerging technologies and the integration of artificial intelligence, we stand on the cusp of a new era in silent communication. As we navigate this brave new world, let us not forget the lessons learned along the way—the importance of observation, stealth, and solidarity in times of need.

In the end, it's not just about speaking without words—it's about building a network of support and understanding that transcends language and borders. So let us continue to learn, adapt, and grow together, ensuring that we are always prepared for whatever challenges may lie ahead.

Book 10: VHF and UHF: Bridging Distances with Technology

Welcome to "Emergency Preparedness and Off-Grid Communication" series: "VHF and UHF: Bridging Distances with Technology." In this comprehensive guide, we're diving headfirst into the world of VHF and UHF communication—the backbone of effective communication in times of crisis.

Picture this: you're in the midst of a disaster, cut off from traditional communication channels. Your cell phone is useless, and you're left wondering how to reach out for help or coordinate with others. This is where VHF and UHF communication come into play. These frequencies offer a lifeline when other methods fail, providing a reliable means of staying connected and informed when it matters most.

Throughout this book, we'll leave no stone unturned as we explore the ins and outs of VHF and UHF communication. From understanding the basics of frequencies and equipment to setting up your station and maximizing your reach with advanced strategies, we've got you covered every step of the way.

But this isn't just about learning the technical aspects of VHF and UHF communication—it's about empowering you to take control of your safety and well-being in any situation. Whether you're a beginner looking to dip your toes into the world of off-grid communication or an experienced enthusiast seeking to expand your skill set, this guide is designed to meet you where you are and take you to the next level.

We will start by looking at VHF and UHF communication—the backbone of off-grid communication in emergencies. You're probably wondering what all the buzz is about, so let's break it down, starting with the basics.

First off, let's talk frequencies and equipment. VHF stands for Very High Frequency, ranging from 30 to 300 MHz, while UHF stands for Ultra High Frequency, ranging from 300 MHz to 3 GHz. These bands offer unique advantages, with VHF being better suited for long-distance communication over open terrain, and UHF excelling in urban and obstructed environments.

Now, onto the equipment. You'll need a VHF/UHF transceiver—a device that can both transmit and receive signals on these frequencies. These come in all shapes and sizes, from handheld radios to mobile units for vehicles, and even base stations for fixed installations. Choose the right equipment based on your needs and budget, ensuring it's rugged and reliable for use in challenging conditions.

But hold on, before you rush out to buy that shiny new radio, let's talk about the capabilities and limitations of VHF/UHF communication. While these frequencies offer excellent line-of-sight communication, they're not without their challenges. They can be affected by terrain, weather, and obstructions like buildings or foliage, which can limit their range and effectiveness.

That said, when used correctly, VHF/UHF communication can be a game-changer in emergencies. It provides a reliable means of communication when other methods fail, allowing you to stay in touch with family, friends, or emergency services when traditional networks are down or overloaded.

Picture this: you're out in the wilderness on a hiking trip when disaster strikes. Your cell phone has no signal, and you're miles away from the nearest town. This is where VHF/UHF communication shines. With the right equipment and know-how, you can reach out for help, coordinate with others in your group, or even relay vital information to rescuers.

In essence, VHF/UHF communication is like having your own private lifeline in emergencies. It gives you the power to stay connected and informed when it matters most, empowering you to take control of your safety and well-being in any situation.

Setting Up Your VHF/UHF Station: A Step-by-Step Guide

Alright, let's roll up our sleeves and get your VHF/UHF station up and running. This step-by-step guide is your ticket to unlocking the power of off-grid communication, so let's dive in.

First things first: selecting the right equipment and antennas. When it comes to VHF/UHF communication, quality gear is key. Look for a reliable transceiver that covers the frequencies you need and has the features you want—whether that's waterproofing for rugged outdoor use or advanced programming capabilities for power users. As for antennas, choose one that matches your intended use case—whether it's a simple whip antenna for handheld radios or a directional Yagi antenna for long-distance communication.

Once you've got your gear sorted, it's time to set up shop. Find a suitable location for your station—one that's free from obstructions and interference—and mount your antenna securely. Whether you're installing it on a rooftop, a mast, or a tripod, make sure it's elevated and clear of obstacles to maximize your signal strength.

When choosing a transceiver, consider factors like frequency coverage, power output, and durability. Opt for a model that covers the VHF and UHF bands relevant to your needs, such as the commonly used 2-meter and 70-centimeter bands. Look for a transceiver with a solid build quality and weather-resistant design, especially if you plan to use it in outdoor or harsh environments. Additionally, consider features like dual watch, memory channels, and programmable squelch to enhance your communication experience.

Now comes the fun part: installing and configuring your VHF/UHF station. Start by connecting your transceiver to your antenna using quality coaxial cable, making sure to use weatherproof connectors to prevent water damage. Then, power up your transceiver and follow the manufacturer's instructions to program your desired frequencies, channels, and settings. Don't forget to set up any necessary repeater offsets or tones for optimal performance.

Once your station is up and running, it's time to fine-tune and optimize its performance. Start by checking your signal strength and quality using a signal meter or the built-in meter on your transceiver. If you're experiencing any issues, troubleshoot common problems like antenna misalignment, interference, or cable damage. Adjust your antenna's position and orientation as needed, and experiment with different settings to find the optimal configuration for your setup.

Finally, remember that practice makes perfect. Spend some time familiarizing yourself with your VHF/UHF station, experimenting with different frequencies, modes, and features. Practice making calls, listening for signals, and adjusting your setup for maximum effectiveness. The more you use your station, the more confident and proficient you'll become in operating it during emergencies.

So there you have it—a step-by-step guide to setting up your VHF/UHF station. With the right equipment, installation know-how, and a bit of practice, you'll be well on your way to harnessing the power of off-grid communication and bridging distances with technology. Now go forth and communicate with confidence!

Maximizing Your Reach: Tips and Tricks for VHF/UHF Communication

Alright, let's get down to some advanced strategies for optimizing your VHF/UHF communication setup. Whether you're aiming to extend your reach, enhance clarity, or navigate challenging conditions, these tips have got you covered.

Fine-tuning your equipment for optimal performance is essential. Ensure your antenna's SWR (Standing Wave Ratio) is properly adjusted to match your frequencies. Use an SWR meter or your transceiver's built-in feature to trim antenna elements until achieving the desired SWR. This adjustment is crucial for maximizing transmission range and minimizing interference.

Repeater usage is another valuable tool for extending communication range. Familiarize yourself with local repeaters, noting their frequencies, offsets, and PL tones. When accessing a repeater, always listen first to avoid interrupting ongoing transmissions. Transmit only after hearing the courtesy tone to maintain communication etiquette and efficiency.

When you're dealing with tough conditions like interference or bad weather, choosing the right frequency is key. Lower frequencies travel farther and can go through obstacles better, while higher frequencies give you clearer signals in open areas. Experiment with frequency selection and antenna configurations to optimize your setup for your specific environment.

Now, let's talk about some advanced tricks that can really make a difference. Ever heard of frequency hopping? It's like constantly changing radio channels while you're talking, making it hard for anyone trying to listen in or mess with your signal. This keeps your messages safe and sound, even when things get chaotic. This method enhances communication reliability by reducing the impact of interference and jamming attempts, ensuring your messages remain secure and intact.

Spread spectrum modulation is another cool technique. It spreads your signal out over a wide range of frequencies, making it tougher for interference or noise to mess things up. By spreading the signal's energy over a wide frequency range, spread spectrum modulation enhances communication resilience and minimizes the effects of narrowband interference or jamming. This comes in handy, especially in crowded cities or places with lots of electronic noise.

When implementing advanced techniques like frequency hopping or spread spectrum modulation, it's essential to ensure compatibility with your equipment and infrastructure. Invest in transceivers and accessories that support these features and familiarize yourself with their operation and configuration settings. Get yourself some gear that's up to the task and get familiar with how it all works. Test things out in different situations to find what works best for you. Experiment in order to optimize performance and reliability in real-world scenarios, adjusting parameters as needed to achieve optimal results.

By using these advanced tricks with your VHF/UHF gear, you'll be ready to tackle anything that comes your way. Whether it's interference, jamming, or just a crowded area, these methods will keep your communication on point. So, play around with different techniques until you find what works best for you, and you'll be ready for whatever comes your way.

In short, to make the most of your VHF/UHF communication setup, focus on getting your gear right, making the most of repeaters, and trying out advanced techniques. With these tips, you'll be ready to handle any communication challenge that comes your way.

Maximising your VHF/UHF communication capabilities involves optimizing equipment performance, leveraging repeaters effectively, and employing advanced communication techniques. With these strategies, you'll be well-prepared to overcome obstacles and communicate effectively in any situation.

Enhancing Communication Security: Encryption and Privacy Measures.

When it comes to emergency communication, ensuring the security of your transmissions is paramount. In this section, we're diving into the world of encryption and privacy measures for VHF/

UHF communication. Let's talk about how you can safeguard your messages and protect sensitive information when the stakes are high.

First off, let's tackle encryption. Encryption is like putting your message in a lockbox before sending it out into the world. It scrambles your message into a jumbled mess that only the intended recipient can unscramble with the right key. There are various encryption methods available for VHF/UHF communication, each offering different levels of security and complexity. From basic codes to advanced cryptographic algorithms, the key is to choose a method that strikes the right balance between security and usability for your specific needs.

Now, onto privacy measures. Just like you wouldn't want strangers listening in on your private conversations, you don't want unauthorized parties intercepting your emergency messages. Implementing privacy measures involves taking steps to prevent eavesdropping and unauthorized access to your communication channels. This might include using secure frequencies, employing frequency hopping techniques, or utilizing privacy codes to restrict access to your transmissions.

When selecting an encryption method, consider factors such as the level of security required, the ease of implementation, and compatibility with your existing equipment.

One commonly used encryption method for VHF/UHF communication is frequency inversion scrambling. This method involves shifting the frequency of the transmitted signal to encode the message, making it unintelligible to anyone without the proper decryption key. Frequency inversion scrambling is relatively simple to implement and offers moderate security against casual eavesdropping. However, it may not be sufficient for highly sensitive communications or in environments where sophisticated adversaries are present.

For higher levels of security, consider more advanced encryption algorithms such as Advanced Encryption Standard (AES) or Data Encryption Standard (DES). These algorithms use complex mathematical algorithms to encrypt messages, providing strong protection against interception and decryption attempts. While implementing AES or DES encryption may require more advanced equipment and technical expertise, the level of security they offer is well worth the investment, particularly for critical communications in high-risk environments.

Ultimately, the key to effective encryption is striking the right balance between security and usability. Choose an encryption method that offers the level of protection appropriate for your communication needs while still allowing for efficient and reliable transmission of messages. By selecting the right encryption method and implementing it effectively, you can ensure that your VHF/UHF communications remain secure and confidential, even in the most challenging circumstances.

But encryption and privacy measures alone aren't enough to ensure secure communication channels. You also need to follow best practices for maintaining the integrity and security of your transmissions. This means regularly updating your encryption keys, keeping your equipment secure from tampering or theft, and being vigilant for any signs of unauthorized access or interference.

Remember, the goal of enhancing communication security isn't just to protect your own information—it's also about safeguarding the integrity of your network and ensuring that critical messages reach their intended recipients without being intercepted or compromised. By understanding encryption methods, implementing privacy measures, and following best practices for maintaining secure communication channels, you can bolster the security of your VHF/UHF transmissions and communicate with confidence in even the most challenging circumstances.

Building Reliable Networks: Establishing VHF/UHF Communities.

Let's look into the critical task of building reliable networks using VHF/UHF communication. In emergencies, having a network of like-minded individuals and groups can mean the difference between safety and vulnerability. Here's how you can establish and maintain robust communication networks with VHF/UHF technology.

Connecting with individuals who share your preparedness mindset is fundamental. Start by reaching out to neighbors, friends, and local organizations interested in emergency readiness. Attend community meetings, participate in online forums dedicated to self-reliance, or engage in workshops focused on disaster preparedness. By forging these connections, you lay the groundwork for a supportive network that can offer assistance when crises arise.

Once you've identified potential network members, it's crucial to organize regular check-ins and drills to ensure network reliability. Set up weekly or monthly check-ins where members can test their equipment, practice communication protocols, and share updates on their preparedness efforts. Conducting drills that simulate various emergency scenarios helps identify weaknesses in your network and allows for proactive adjustments.

These drills simulate real-life scenarios, allowing network members to practice their communication protocols and refine their response procedures. For example, you could organize a drill where members simulate a power outage or a severe weather event and practice using VHF/UHF radios to relay important information, such as their status, location, and any immediate needs. By participating in these drills, network members can familiarize themselves with the equipment, test the effectiveness of their communication strategies, and identify areas for improvement. Additionally, these exercises help build trust and camaraderie among network members.

Sharing resources and information within the VHF/UHF community is another vital aspect of building reliable networks. Establish channels for disseminating relevant information, such as local weather updates, emergency alerts, or resource availability. Utilize online platforms, social media groups, or dedicated communication channels to ensure swift and efficient information sharing among network members.

For instance, consider creating a dedicated online forum where network members can post updates on their preparedness activities, share tips and tricks, and ask for assistance when needed. Additionally, designate specific frequencies or communication channels for emergency use, ensuring that all network members are aware of these channels and know how to access them during crises.

By fostering a sense of community and collaboration, you can harness the power of VHF/UHF communication to build resilient networks that support each other during times of adversity. Together, you and your network members can overcome challenges, share resources, and ultimately increase your collective chances of survival in emergency situations.

Mobile Operations: Communication on the Move

Communication is your lifeline, whether you're on the move or stationed in one place. In this section, we'll explore how to maintain effective VHF/UHF communication while you're on the go, ensuring that you stay connected no matter where your adventures take you.

When it comes to mobile operations, setting up your VHF/UHF station correctly is key. You need equipment that's compact, reliable, and easy to transport. Look for handheld radios or mobile transceivers specifically designed for on-the-go use. These devices are rugged enough to withstand the rigors of travel while providing the performance you need to stay connected.

Once you've got your equipment sorted, it's time to strategize. Plan your communication routes in advance, taking into account potential obstacles such as terrain features or urban congestion. Consider alternative routes and backup plans in case your primary route is compromised.

Imagine you're embarking on a backcountry hiking trip with a group of friends, each equipped with handheld VHF/UHF radios. Before setting out, you designate a leader responsible for communication and assign each member a specific channel for check-ins. As you navigate the trail, your group encounters a sudden downpour, which disrupts your visibility and makes it challenging to maintain visual contact.

In this scenario, your VHF/UHF radios become invaluable tools for staying connected and ensuring everyone's safety. Using the designated channels, you conduct regular check-ins to monitor each member's progress and address any concerns that arise. When one member encounters a fallen tree blocking the trail, they immediately relay the information over the radio, allowing the group to reroute and avoid the obstacle safely. Throughout the hike, clear communication via VHF/UHF radios enables your group to overcome challenges, navigate obstacles, and ultimately reach your destination safely.

In remote areas where infrastructure is scarce, maintaining communication can be challenging. However, with the right strategies, you can overcome these obstacles. Utilize repeaters strategically to extend your range, and consider investing in directional antennas to boost signal strength in areas with poor coverage.

Optimizing your mobile antenna setup is crucial for maintaining clear communication while on the move. Choose antennas that are compact and lightweight without sacrificing performance. Magnetic-mount antennas are a popular choice for vehicle-mounted setups, offering easy installation and removal without causing damage to your vehicle's exterior.

Consider the height and location of your antenna for maximum effectiveness. Mounting your antenna on the roof of your vehicle or on a mast can significantly improve signal propagation, especially in hilly or wooded terrain.

By selecting the right equipment, strategizing your communication routes, and optimizing your antenna setup, you can ensure reliable communication on the move, keeping you connected and informed wherever your adventures take you.

Emergency Communication Protocols: Responding to Crisis Situations

When disaster strikes, effective communication becomes the cornerstone of survival. Establishing clear and efficient emergency communication protocols is essential for navigating crisis situations with confidence and ensuring the safety of yourself and those around you.

First and foremost, take the time to develop comprehensive protocols for emergency communication procedures. This includes defining roles and responsibilities within your group or community, establishing designated communication channels and frequencies, and outlining procedures for initiating and responding to distress signals. Assign specific tasks to individuals based on their skills and expertise, ensuring that everyone knows their role in the event of an emergency.

Coordination with local authorities and emergency responders is another crucial aspect of effective crisis communication. Familiarize yourself with the emergency communication systems and protocols used in your area, and establish lines of communication with relevant authorities, such as law enforcement agencies, fire departments, and medical services. Share pertinent information about your location, situation, and needs, and follow their instructions and guidance to ensure a coordinated response.

Maintaining calm and clear communication during high-stress situations is easier said than done, but it's essential for effective decision-making and problem-solving. Practice active listening, speak calmly and clearly, and avoid unnecessary jargon or technical language that may cause confusion. Keep your messages concise and to the point, focusing on conveying essential information while remaining open to feedback and input from others.

In times of crisis, every second counts, and clear communication can mean the difference between life and death. By establishing robust emergency communication protocols, coordinating with local authorities and emergency responders, and maintaining calm and clear communication during high-stress situations, you can ensure that you and your community are prepared to face whatever challenges come your way.

One practical example to enhance emergency communication protocols is the implementation of a designated "check-in" system during crisis situations. Here's how it works:

1. Designate a specific time interval for check-ins, such as every hour or every half-hour, depending on the urgency of the situation.
2. Assign individuals within your group or community to take on the role of check-in coordinators. These individuals will be responsible for initiating and overseeing the check-in process.
3. Establish predetermined communication channels and frequencies for check-ins, ensuring that everyone knows where and how to report their status.
4. At the designated check-in times, each member of the group or community will be required to report their status to the check-in coordinator. This can be done via radio, phone, or other communication devices.
5. The check-in coordinator will compile the information received from each member and report back to the group, providing an update on the status of everyone involved.
6. If a member fails to check-in at the specified time, the check-in coordinator will follow up with them to ensure their safety and well-being. If contact cannot be established, appropriate action will be taken based on the established emergency protocols.

By implementing a structured check-in system, you can ensure that everyone in your group or community remains accounted for and that vital information is communicated effectively during crisis situations. This simple yet effective protocol can help maintain order, facilitate coordination, and ensure the safety of all involved.

Antenna Design and Deployment: Maximizing Signal Strength

When it comes to VHF/UHF communication, your antenna plays a pivotal role in ensuring reliable signal transmission. Let's dive into the fundamentals of antenna design and deployment to help you maximize your signal strength and maintain clear communication, no matter the circumstances.

Understanding antenna design principles is crucial for optimizing your VHF/UHF setup. Antennas are essentially conductors designed to transmit and receive electromagnetic waves efficiently. The key factors to consider include antenna type, polarization, gain, and radiation pattern. For VHF/UHF frequencies, common antenna types include dipole, Yagi, and log-periodic antennas, each with its own advantages and applications. By choosing the right antenna type based on your communication needs and operating environment, you can significantly improve signal reception and transmission.

Once you've selected the appropriate antenna type, it's time to build and deploy it effectively. Building your antenna may require some DIY skills, but it's a worthwhile investment in optimizing your communication setup. Ensure that you follow the manufacturer's instructions or reputable guides closely to construct your antenna correctly. Pay attention to details such as element length, spacing, and orientation to achieve optimal performance.

Proper placement and alignment can make a significant difference in signal strength and clarity. Position your antenna in an elevated location free from obstructions to maximize line-of-sight transmission. Avoid placing it near large metal objects or other sources of interference that could degrade signal quality. Additionally, consider using a mast or tower to raise your antenna above surrounding obstacles for improved coverage and range.

When it comes to antenna optimization, how and where you position your gear can spell the difference between a crystal-clear connection and a frustrating jumble of static. Picture this: You're out in the wild, part of a search and rescue crew tackling tough terrain. Dense forests and towering hills loom around you, threatening to block your signals. Here's where smart antenna placement becomes your best ally. Strap your antenna onto a sturdy mast or portable tower, lifting it high above the foliage and obstacles. This simple move boosts your line-of-sight transmission, giving your signal the edge it needs to reach further and clearer. Don't forget to aim for directional antennas with high gain—they help focus your signals where they're needed most, even in the toughest of landscapes.

Now, let's talk interference—a communication buzzkill you want to avoid at all costs. Interference can throw a wrench in your plans, disrupting signals and making your message sound like a garbled mess. That's why it's vital to root out potential sources of interference and neutralize them. Tinker with your antenna's placement and angle to dodge multipath interference, caused by pesky signal reflections bouncing off surfaces. Get savvy with polarization matching and antenna polarization diversity techniques—they're your secret weapons against interference headaches.

To make the most of your VHF/UHF setup, you need to speak the language of antennas fluently. Start by grasping the basics of antenna design and selecting the right type to suit your needs. Then, it's all about the setup game. Build and deploy your antenna with care, paying close attention to placement and alignment. When you've got that down, interference won't stand a chance, and your communication game will be on point, no matter the conditions.

Power Management: Ensuring Long-Term Communication Reliability

When it comes to keeping your VHF/UHF communication up and running, power is paramount. Let's dive into how you can ensure long-term reliability through sustainable power solutions.

For your VHF/UHF stations, you've got a few options on the table when it comes to power sources. Batteries are a classic choice, offering portability and ease of use. Opt for high-capacity batteries like lithium-ion or nickel-metal hydride to keep your transmissions humming for extended periods. Alternatively, solar panels provide a renewable energy source, ideal for off-grid setups or prolonged operations. Harnessing the power of the sun keeps your communication going strong, even when traditional power sources are scarce.

Now, let's focus on making the most of your power resources. Implementing energy-efficient practices is key to maximizing battery life and minimizing downtime. Start by optimizing your equipment settings to strike the right balance between performance and power consumption. Dimming displays, reducing transmission power, and implementing sleep modes are all effective ways to stretch your battery's lifespan. Remember, every watt saved is a step towards long-term communication reliability.

But what happens when the lights go out? That's where backup power options come into play. Whether it's a sudden blackout or a prolonged emergency situation, having a backup power plan ensures you're never left in the dark. Consider investing in backup batteries or generators to keep your communication lines open when the grid goes down. Additionally, explore alternative power sources like hand-cranked chargers or portable fuel cells for added peace of mind.

In summary, sustainable power management is the cornerstone of long-term communication reliability. By choosing the right power sources, implementing energy-efficient practices, and having backup power options in place, you can ensure your VHF/UHF communication remains operational when it matters most. So, don't wait until the lights flicker—take charge of your power management strategy today and stay connected no matter what the future holds.

Tips for Sustainable Power Management:

1. Monitor battery levels regularly and recharge as needed.
2. Invest in energy-efficient equipment to minimize power consumption.
3. Utilize power-saving features and modes on devices.
4. Keep backup power sources fully charged and ready.
5. Integrate renewable energy sources like solar panels.
6. Rotate and test backup batteries periodically.
7. Explore hybrid power solutions for increased reliability.
8. Educate yourself and your team on power management.
9. Store spare batteries and power supplies properly.
10. Regularly review and update your power management strategy.

Future Trends and Innovations: Advancing VHF/UHF Communication Technology.

Let's take a peek into the future of VHF/UHF communication. Exciting stuff ahead, trust me!

First of all, we've got Software-Defined Radios (SDRs) on the rise. Imagine having a radio that can adapt to different frequencies and modes in a snap. With an SDR in your kit, you're like a communication wizard, ready to tackle any situation that comes your way.

And here's the scoop on digital voice modes: they're the future of crystal-clear communication.

Technologies like Digital Mobile Radio (DMR) and Project 25 (P25) deliver top-notch audio quality, even when the going gets tough. Say goodbye to fuzzy signals and hello to crisp, clear communication, no matter what.

Now, let's chat about artificial intelligence (AI). Yes, AI is stepping into the world of VHF/UHF communication, helping us optimize our systems on the fly. It's like having a smart assistant fine-tuning your communication protocols in real-time, making sure your messages get through loud and clear, no matter the conditions.

And finally, let's take a moment to ponder the evolving role of VHF/UHF communication in our rapidly changing world. It's not just about voice communication anymore. These days, VHF/UHF systems are like Swiss Army knives, packing a ton of functionality into one neat package.

So, there you have it – the future of VHF/UHF communication in a nutshell. Exciting times ahead, right? With all these cool advancements on the horizon, we're gearing up for a whole new era of off-grid communication. So, buckle up and get ready for the ride – it's going to be one heck of an adventure!

As we come to the end of our exploration into VHF/UHF communication, it's clear that these frequencies offer valuable tools for bolstering our preparedness and resilience in times of crisis. From understanding the basics of frequencies and equipment to mastering advanced strategies for extending reach and enhancing security, we've covered all the essentials. By tapping into the potential of VHF/UHF communication, we empower ourselves to stay connected, informed, and secure during emergencies.

Throughout our journey, we've focused on practical advice and hands-on guidance, ensuring that you're equipped with the knowledge and tools needed to set up reliable communication stations, establish robust networks, and navigate challenging situations with confidence. Whether it's deploying mobile operations, implementing emergency communication protocols, or optimizing signal transmission through effective antenna design, we've provided actionable steps to enhance your preparedness.

Looking ahead, the future of VHF/UHF communication holds exciting prospects with emerging technologies such as Software-Defined Radios, mesh networking, and artificial intelligence. These innovations promise even greater connectivity and resilience, enabling us to adapt and thrive in an ever-changing world. By staying informed and adaptable, we can embrace these advancements and continue our journey towards greater self-reliance and readiness in the face of uncertainty.

Download Your Fantastic Bonus Scanning This QR Code or Follow This Link:

https://book-bonus.com/emergency-preparedness-and-off-grid-communication/

BONUS #1: The Communication Signal Cheat Sheet

Why rely on fragile grid systems when you can master the art of off-grid communication? Discover top-secret comms tactics including Morse code, hand signals, radio codes, and life-saving whistle signals. This cheat sheet ensures you stay connected, command operations, and maintain safety in the most critical moments—making you the master of any situation. Essential, robust knowledge every prepper must have to thrive in the wild or during chaos!

BONUS #2: Survival Checklist and Gear Guide

What if you could be prepared for any crisis at a moment's notice? Unlock our exclusive Survival Checklist and Gear Guide to equip yourself with the essential tools and knowledge needed to handle any emergency. From the must-have survival gadgets to the basics for daily resilience, this guide ensures you're never caught unprepared. Dive into our meticulously crafted list and transform your readiness into a fine-tuned art. Essential for every survivalist looking to fortify their defenses against the unknown!

BONUS #3: Survival Gardening and Seed Saving Guide

Why depend on supply chains when you can grow your own survival garden? This guide plunges you into the essentials of sustainable living with hands-on techniques for cultivating a thriving garden and saving seeds for the future. Learn the secrets to developing a robust garden ecosystem that not only sustains but flourishes through any crisis. Essential for preppers who want to achieve true food independence and resilience!

www.ingramcontent.com/pod-product-compliance
Lightning Source LLC
Chambersburg PA
CBHW062217220526
45471CB00009B/3244